1st and 2nd International Conference on High Power Impulse Magnetron Sputtering

(HIPIMS 2010 and 2011)

IOP Conference Series: Materials Science and Engineering Volume 39

Sheffield, United Kingdom
6 – 7 July 2010

Braunschweig, Germany
28 – 29 June 2011

ISBN: 978-1-63439-850-3
ISSN: 1757-8981

1st and 2nd International Conference on High Power Impulse Magnetron Sputtering

(HIPIMS 2010 and 2011)

IOP Conference Series: Materials Science and Engineering Volume 39

Sheffield, United Kingdom
6 – 7 July 2010

Braunschweig, Germany
28 – 29 June 2011

ISBN: 978-1-63439-850-3
ISSN: 1757-8981

IOP Institute of Physics

Printed from e-media with permission by:

Curran Associates, Inc.
57 Morehouse Lane
Red Hook, NY 12571

Some format issues inherent in the e-media version may also appear in this print version.

Printed by Curran Associates, Inc. (2015)

For permission requests, please contact the Institute of Physics
at the address below.

Institute of Physics
Dirac House, Temple Back
Bristol BS1 6BE UK

Phone: 44 1 17 929 7481
Fax: 44 1 17 920 0979

techtracking@iop.org

Additional copies of this publication are available from:

Curran Associates, Inc.
57 Morehouse Lane
Red Hook, NY 12571 USA
Phone: 845-758-0400
Fax: 845-758-2634
Email: curran@proceedings.com
Web: www.proceedings.com

TABLE OF CONTENTS

PREFACE

Thin film technology and surface engineering are nowadays key components for numerous innovative products like efficient windows, flat screens, sensors or hard coatings used in tool coating and automotive applications, as well as products for everyday life. In line with the demands of surface technology, coating technology is also evolving and improving. The latest major technology jump was the introduction of pulse technology in physical vapor deposition. High power impulse magnetron sputtering is the most recent development of pulse sputtering. After approximately a decade of intense academic investigation and development we observe today a transfer of this new technology towards industrial processes.

As well as several international activities the international conference on fundamentals and applications of HIPIMS continues the success story of the HIPIMS days, initiated in 2004 at Sheffield Hallam University, UK. Becoming the only international conference especially dedicated to HIPIMS the HIPIMS conference is a venue for industrial and academic exchange on the latest developments in this fast evolving new technology. As a joint undertaking of Sheffield Hallam University SHU, Network of Competence for Industrial Plasma Surface Technology INPLAS and Fraunhofer Institute for Surface Engineering and Thin Films IST the HIPIMS conference was launched in 2010 in Sheffield, UK. With 120 delegates the impact of the new conference was underlined. The growing importance of HIPIMS technology was connected with a growth by nearly 35% to 160 participants in 2011 at the second HIPIMS conference in Braunschweig, DE. The participants were made up of equal numbers from research and development (university and research institutes) and industry. Being a global conference representatives from 25 different countries from all continents attended.

The HIPIMS conference is also in joint collaboration with the COST Action MP0804 Highly Ionized Pulse Plasma Processes (www.hipp-cost.eu). COST (European Cooperation in Science and Technology) is one of the longest-running European frameworks supporting cooperation among scientists and researchers across Europe (www.cost.eu). The COST Action MP0804 HIPP processes focuses on the fundamentals and the industrial implementation of highly ionized pulse plasmas, where HIPIMS is the most prominent and most mature technology, today.

Over 50 high level contributions, divided in 37 oral and 14 poster presentations were highly appreciated by the professional audience. The message from 2011 was that HIPIMS technology has now reached industry. In the opening session of the conference representatives from different companies reported on the latest developments in industrialization. Using HIPIMS technology, the lifetime of mills using a state of the art coating can be extended by 50%. Comparable deposition rates for coating cutting inserts on the different faces are reported. The ice-free window for automotive application is one solution just becoming available by HIPIMS technology. The talks from international experts covered a range from fundamental physics, experimental investigations, theoretically modeling to several applications and made the international conference on fundamentals and applications a success story to be continued in the following years.

Arutiun Ehiasarian and Ralf Bandorf
(Conference Chairmen of HIPIMS 2010 and 2011, respectively)

Organising Committee (2010 and 2011) – Affiliations
Professor Dr Papken Hovsepian (Sheffield Hallam University, Nanotechnology Center for PVD Research, UK) Professor Dr Günter Bräuer (Fraunhofer Institute for Surface Engineering and Thin Films IST/ Network of Competence INPLAS, Braunschweig, DE) Professor Dr Arutiun P. Ehiasarian (Sheffield Hallam University, Materials Research Institute, UK) Dr Ralf Bandorf (Fraunhofer Institute for Surface Engineering and Thin Films IST, Braunschweig, DE)

Main Sponsor
Society of Vacuum Coaters SVC
Albuquerque, New Mexico, USA

Sponsors
Ionbond Netherlands BV
Venlo, NL
Hauzer Techno Coating BV
Venlo, NL
Hüttinger Elektronik GmbH + Co. KG
Freiburg, DE

Conference Photos

Attendees HIPIMS Conference 2010, Sheffield, UK

Attendees HIPIMS Conference 2011, Braunschweig, Germany

IOP Conf. Series: Materials Science and Engineering **39** (2012) 012001 doi:10.1088/1757-899X/39/1/012001

Titanium aluminum nitride sputtered by HIPIMS

Juergen Weichart, Markus Lechthaler

OC Oerlikon Balzers Ltd., LI-9496 Balzers, Liechtenstein

Abstract. TiAlN was sputtered reactively by HIPIMS in the target compositions Ti/Al 33/67 and 50/50 using a modified OC Oerlikon Balzers INNOVA coating equipment. The resulting film properties like deposition rate, surface roughness, hardness, Young's modulus, wear, and film stress were analyzed as function of the nitrogen gas flow, pressure, target-substrate distance, and substrate bias. Furthermore, the films were characterized by X-ray diffraction and secondary electron microscopy of the cross section and the surface appearance. The process characteristics and film properties were compared with pulsed DC sputtering under the same conditions.

Introduction

$Ti_xAl_{(1-x)}N$ is well known as hard and wear resistant coating providing a good resistance to high working temperature. It is typically sputtered or arc-evaporated in nitrogen atmosphere from the metal compound target. For x: 0.5 the highest as-deposited hardness of $Ti_xAl_{(1-x)}N$ has been shown [1], whereas less columnar and more fine grained material is found for x>0.5 [1] providing a longer tool life at high working temperatures for the meta-stable composition of x: 0.33 [2].

High Power Impulse Magnetron Sputtering (HIPIMS) is a new process working with pulsed DC at a very low duty cycle of less than 10% and high current of 300 A up to more than 1000 A or current densities of typically 1 A/cm^2 leading to a high ionization of the sputtered material [3, 4]. By this technology film properties comparable to arc-evaporated TiAlN are expected, but with a much smoother surface which is free of macroparticles [5].

Experimental and process setup

The HIPIMS generator Huettinger TruPlasma Highpulse has been installed as a replacement on the rectangular DC magnetron source AKQ525 by OC Oerlikon Balzers with a target size of 780cm^2 without any change of the source or magnet design on the industrial tool coater INNOVA. Two target types, TiAl 50/50 at% produced by casting, and powder metallurgically prepared TiAl 33/67 at% have been run in this study. A stable and smooth HIPIMS discharge could be run up to high reactive gas flows of N_2. In front of the HIPIMS cathode a single rotating tree was mounted providing different target substrate distance (TSD) position between 55 and 145mm. Prior to the deposition the substrate is heated at 450°C for 30 min and etched for 30 min in an Ar discharge with a substrate bias of -170 V. The HIPIMS generator is running with a pulse length of 200 µsec, peak currents of 460 to 600 A at a repetition frequency of 400 Hz and an average power of 15 kW. Ar and N_2 flow rates can be adjusted in the range between 50 and 400 sccm resulting in a total process pressure between 0.2 and 1 Pa. On the substrate carousel a bias voltage between -25 V and -130 V is applied. The deposition rates were ranging from 17 to 36 nm/min for coatings deposited from TiAl 50/50 at% targets (here named TiAlN) and 26 to 50 nm/min for coatings deposited from TiAlN 33/67 at% targets (here named AlTiN).
For comparison with the DC case the Advanced Energy PinnaclePlus was installed using a pulse frequency 20 kHz, an off-time of 0.4 µsec and an average power of 10 kW. The lower average power

-----
Corresponding author.
E-mail address: Juergen.weichart@oerlikon.com (J Weichart)

has been selected to work at similar deposition rates as for the HIPIMS case of 30 to 65 nm/min. for AlTiN.

Fig 1 shows the average source current and average bias current for HIPIMS and DC of reactively sputtered AlTiN adjusted to the same deposition rate as function of the N_2 flow. The source current indicates that the transition to the reactive mode happens at the same amount of N_2 for HIPIMS as for DC. It was observed that the mean bias current to the substrate cage – here set to a voltage of -50 V - is typically a factor of 2.5 to 4 higher than with DC, indicating a higher amount of metal and Argon ions. Optical emission spectroscopy (OES) of the discharge watched through a window opposite to the target on the INNOVA has shown a high fraction of ionized metal of Ti and Al from the TiAl target. The first analysis of these OES data – which are not presented here - confirm similar findings as presented in the literature [6, 7].

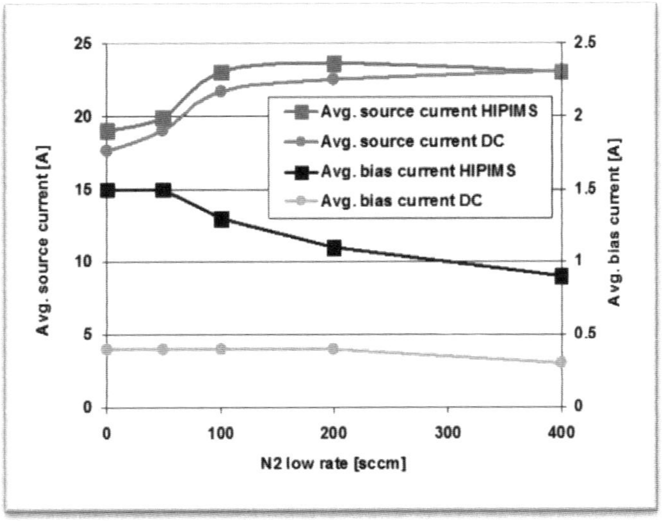

Figure 1. Comparison of the average source current and average bias current for HIPIMS and DC of reactively sputtered AlTiN at the same deposition rate (HIPIMS 15kW, DC 10kW).

Results

It was found that the best film properties could be achieved for low nitrogen flow rates just enough to achieve stoichiometric films. For higher N_2 flow a slight decrease of the hardness, the Young's modulus and the deposition rate is observed. A low Argon pressure is preferred to get a uniform cubic phase of <200> orientation for TiAlN even at higher target substrate distances, as shown by X-ray detection (XRD) analysis.

Significant differences in the morphology of HIPIMS vs. DC can be observed in the SEM cross section (Fig. 2). Where DC sputtering gives columnar growth and a porous film up to relatively high substrate bias voltages (-100 V), the HIPIMS layers deposited at the same conditions - namely the same deposition rate - reveal a very fine grained structure even for low substrate bias voltages of -25 V. In order to achieve good hardness and adhesion however a substrate bias of -80 V is recommended. All analyzed DC samples have shown a low hardness and predominantly hexagonal phase and almost no cubic phase in the XRD analysis. A low compressive stress of not more than -1.4 MPa was achieved for the DC coatings even for high bias voltages like -100 V.

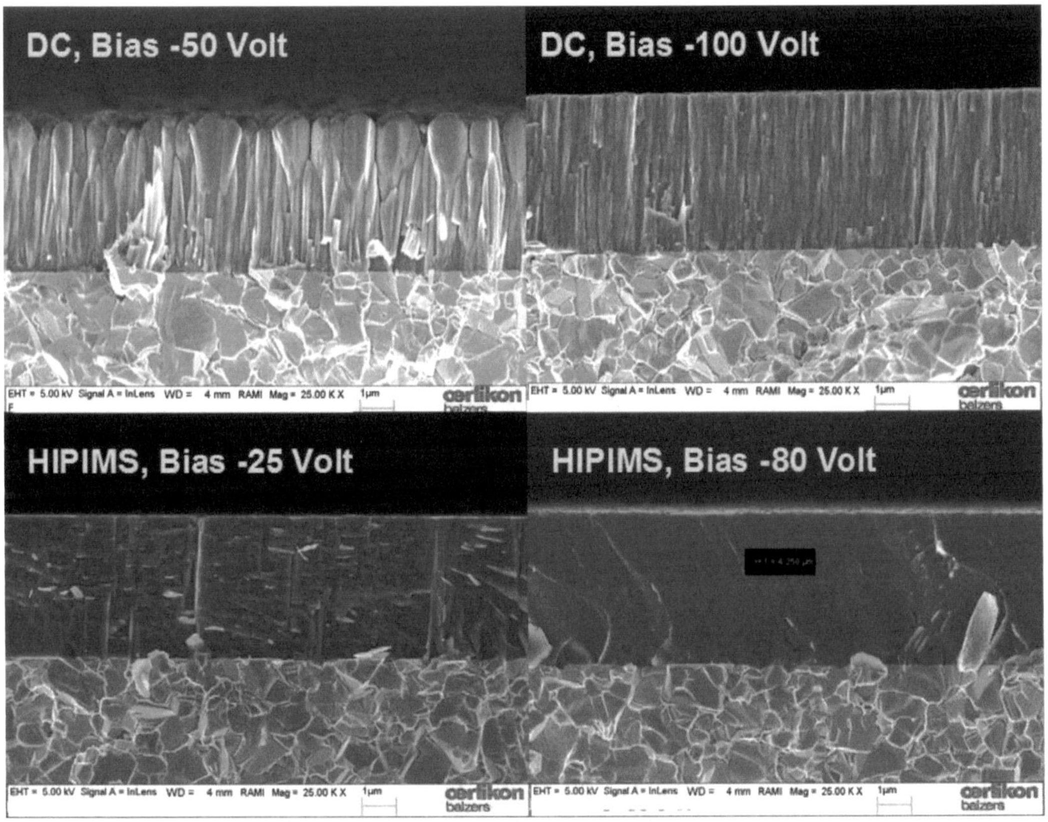

Figure 2. Comparison of SEM cross sections of AlTiN films sputtered by HIPIMS and DC at different DC bias voltages.

The process characteristics and properties of HIPIMS and DC sputtered TiAlN and AlTiN films are compared in table 1. The significant differences in hardness, Young's modulus, and stress are obvious while the surface roughnesses are good for HIPIMS and DC layers.

Table 1. Comparing TiAlN processes deposited by HIPIMS and DC.

	TiAlN 33/66	**TiAlN 50/50**	TiAlN 33/66 DC
Rate [µm/h]	2.5 µm	1.8 µm	2.1 µm/h
Average power [kW]	15	15	10
Hardness [HV]	2400	3000	1400
Young´s Mod [GPa]	300	450	220
Roughness Ra [µm/h]	0.05	0.05	0.04
Stress [GPa]	-2 … -3.6	-3.5	- 1.4

Some first preliminary application tests were performed with HIPIMS TiAlN with 3 flute carbide endmills having a diameter of 8 mm in DIN 1.1191 (190 HB)) using a cutting speed of v_c = 290 mm, a depth of cut of a_p =10 mm and an engagement of a_e = 0.5 mm and a feed rate of f_t = 0.10 mm. They showed a performance in terms of tool life, which was exceeded for more than 50% the tool life of a commercially available sputtered TiAlN coating.

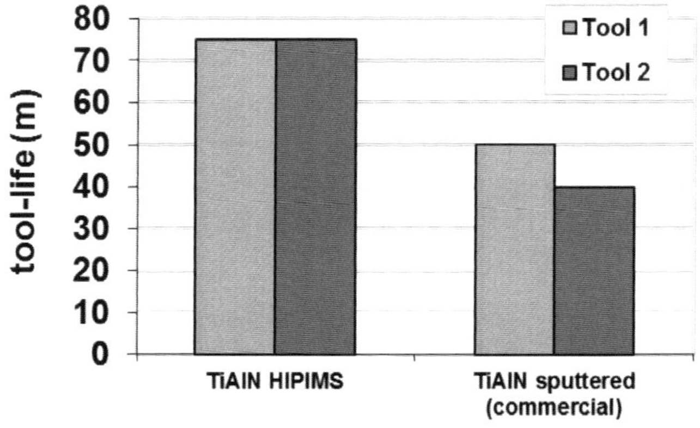

Figure 3. Milling tool life in a finishing application of TiAlN HIPIMS compared with a commercially available sputtered TiAlN coating. Two tools were tested of each variant.

Conclusions

HIPIMS has been successfully applied to deposit $Ti_xAl_{1-x}N$ on an industrial machine (INNOVA). The process conditions were investigated to adjust morphology, structure and mechanical film properties for TiAlN and AlTiN. The results were compared with those of DC sputtering for the same process environment and deposition rate showing very poor performance. The HIPIMS coatings revealed very promising mechanical properties in comparison to DC sputtered TiAlN films and a significantly improved application performance of tools even in the first tests without any further optimization. Compared to the arc evaporation technology the main disadvantage of HIPIMS is the low growth rate. The main advantage of HIPIMS coatings however is their low roughness compared to arc-evaporated layers.

Literature

[1] A. Kimura, et al., Metastable $Ti_{1-x}Al_xN$ films with different Al content, Journal of Mat. Sci. Lett. 19 (2000) 601– 602
[2] A. Hörling, et al., Mechanical properties and machining performance of $Ti_xAl_{(1-x)}N$-coated cutting tools, Surface & Coatings Technology 191 (2005) 384-392
[3] V. Kouznetsov, et al., A novel pulsed magnetron sputter technique utilizing very high target power densities, Surf. Coat. Technol., 122, 290, 1999
[4] U. Helmersson, et al., Ionized Physical Vapor Deposition (IPVD): A Review of Techn. and Appl., Thin Solid Films, 513, (2006) 1-24.
[5] J. Böhlmark et al., Reactive Film Growth of TiN by Using High Power Impulse Magnetron Sputtering (HIPIMS), Society of Vacuum Coaters 505/856-7188, 49th Annual Technical Conference Proceedings (2006) ISSN 0737-5921
[6] A.P. Ehiasarian, Industrial Size High Power Impulse Magnetron Sputtering, 47th Annual Technical Conference Proceedings Society of Vacuum Coaters (2004) 505/856-7188
[7] J. Böhlmark, et al., Ionization of sputtered metals in high power pulsed magnetron sputtering, J. Vac. Sci. Technol. A 23(1) 2005, 18-22.

IOP Conf. Series: Materials Science and Engineering **39** (2012) 012002 doi:10.1088/1757-899X/39/1/012002

Investigations on tailoring the deposition conditions in HIPIMS by varying the pulse durations and the argon partial pressure

M Balzer, M Fenker

FEM – Research institute for precious metals and metals chemistry
Dipl.-Ing. (FH) Martin Balzer, Department: POT-MPh,
Katharinenstrasse 17, 73525 Schwäbisch Gmünd, Germany

Abstract. In most cases HIPIMS is used to get the highest possible ionisation of the deposition particles, which is realised by pulse durations with short on- and very long off-times. These conditions are combined with a more or less pronounced decrease in deposition rate. In this work the pulse configuration has been varied. A three dimensional matrix of parameters was spanned, made of 3 on- and 3 off-times at 4 argon partial pressures. The average power was kept constant and the data achieved were additionally compared to DC-magnetron sputtering. The experiments were carried out using 50 mm diameter targets made of Ti, powered by a MELEC SPIK1000A pulser unit. The deposition rate was measured by quartz microbalance mounted in front of the target. Peak current density and target voltage were recorded and time averaged optical emission spectroscopy (t.a.OES) measurements provided information about the ionisation conditions in the plasma. The results of the data analysis provide a coherent overview of the impact of the HIPIMS parameters as well as of their complex interrelations.

1. Introduction

Since its first introduction in 1999 [1] high power impulse magnetron sputtering (HIPIMS) has proved to be a promising method for depositing hard and dense coatings by producing highly ionized metal vapour. This high ionization is realized by supplying short high power pulses instead of a constant voltage to a magnetron sputtering cathode. A lot of papers have been published on HIPIMS to date, which mostly determined a more or less pronounced decrease of the deposition rate compared to classical DC magnetron sputtering (DC-MS) as the major drawback of this technology.

The idea of the presented measurements was to obtain basic data on HIPIMS plasma and the influence of the process parameters for sputtering of titanium. Essential dependencies had to be identified in order to find a suitable compromise between the ionization of sputtered particles and deposition rate. Since the first application of HIPIMS after this study was the deposition of Ti_3SiC_2 MAX-phase, the major intention was to gain a high amount of single ionized Ti-ions together with a low amount of double ionized ones, which should prevent a destruction of the growing nanolayer.

Corresponding author.
E-mail address: balzer@fem-online.de (M Balzer); phone: +49 (0)7171 - 1006-401; fax: +49 (0)7171 - 1006-900

2. Experimental details

A Ti-target with a size of Ø 50 mm was operated by a MELEC SPIK 1000A pulser providing a maximum peak current of 300 A and a maximum voltage of -1000 V. All measurements were done using power regulation mode and an average target power P_{av} of 250 W (deviations due to voltage limitation as indicated). No other ion source and no superimposed DC- or MF power was used. Three parameters have been varied: The pulse time t_{on} (8 / 15 / 30 µs), the cycle duration T (500 / 1000 / 2000 µs, table 1) as well as the argon pressure p_{Ar} (0.25 / 0.4 / 0.6 / 1.0 Pa). This resulted in a matrix containing 36 sets of parameters. These parameters lie in a similar range as those used by Konstantinidis in 2006 [2]. Pulse shapes were recorded with a Picoscope 3204 PC oscilloscope with "persistence mode" (which keeps elapsed pulses visible for some time, figure 1) enabling reliable measurements of the peak current. Deposition rates were measured by quartz micro balance (INFICON IC 6000, Figure 2), mounted at a distance of about 4.5 cm from target surface. Time averaged optical emission spectroscopy (t.a.OES) measurements were carried out by using an Ocean Optics HR2000CG-UV-NIR with an optical resolution of 1 nm.

Table 1. Duty cylces [%], $T = t_{on} + t_{off}$.

T \ t_{on}	**8 µs**	**15 µs**	**30 µs**
500 µs	1.6	3	6
1000 µs	0.8	1.5	3
2000 µs	0.4	0.75	1.5

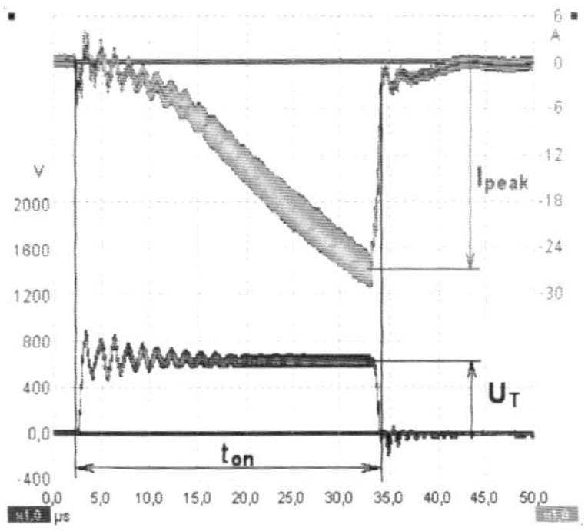

Figure 1. Typical pulse shape in persistence mode (t_{on}=30 µs).

Figure 2. Sputtering arr. with quartz micro balance.

3. Results and discussion

3.1. *Target voltage and peak current density.*

Analysing the data achieved it turned out that the peak current density i_{peak} can be adjusted in a wide range by varying the pulse configurations as well as the Ar-pressure (Figure 3). The highest peak current densities are found for short t_{on} and long T and at low p_{Ar} with an absolute maximum value of 4.5 A/cm^2 for t_{on} = 15 µs, T = 2000 µs at p_{Ar}= 0.4 Pa.

Figure 4 shows that with longer T, shorter t_{on} and lower p_{Ar} the target voltage U_T increases and finally approaches -1000 V, which is the limit of the pulser. Therefore P_{av} is less than 250 W for some sets of parameters. The increase of U_T with longer T and shorter t_{on} can be attributed to the automatic power regulation, which is realized by an automatic target voltage adjustment. Comparing the parameters it can be seen that the highest peak current densities i_{peak} basically coincide with the highest target voltages.

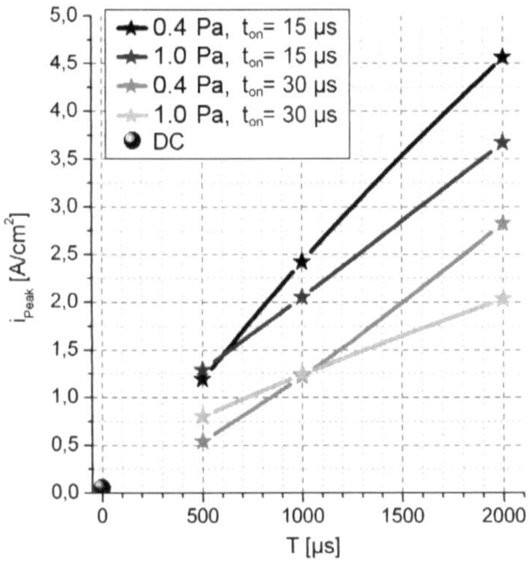

Figure 3. i_{peak} vs. T at two different t_{on} and two p_{Ar}

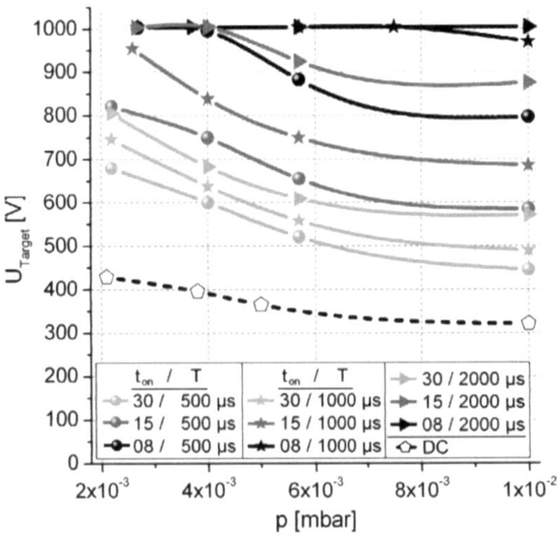

Figure 4. U_T vs. p_{Ar} at different pulse configurations

3.2. Deposition rate

Figures 5 and 6 each show a survey of deposition rates at different pulse configurations versus their belonging peak current densities i_{peak}. It's very interesting to always find an exponential decrease of the rate with increasing i_{peak} at one Ar-pressure chosen. This type of decrease seems to be strictly valid for all pulse configurations, i.e. there is no way to leave the path of the curve except changing the Ar-pressure. Furthermore it becomes obvious, that one and the same set of deposition rate and i_{peak} can be realised using different pulse configurations. The trend of a more or less pronounced decrease of deposition rate (or sputtering efficiency) with increasing target current- or target power density during the pulse (t_{on}) is supported e.g. by [3-5] for several materials. This trend in turn is explained by an increase of the ionization of Ti, which is discussed in the next section.

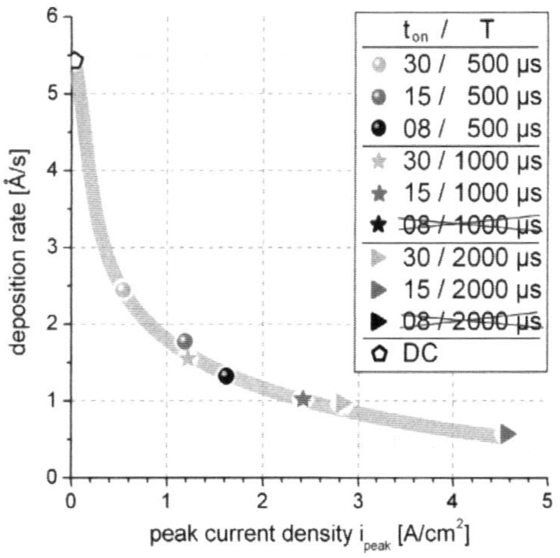

Figure 5. Deposition rate of Ti vs. i_{peak} for different pulse times at p_{Ar} = 0.4 Pa (data points with P_{av} < 250 W not included).

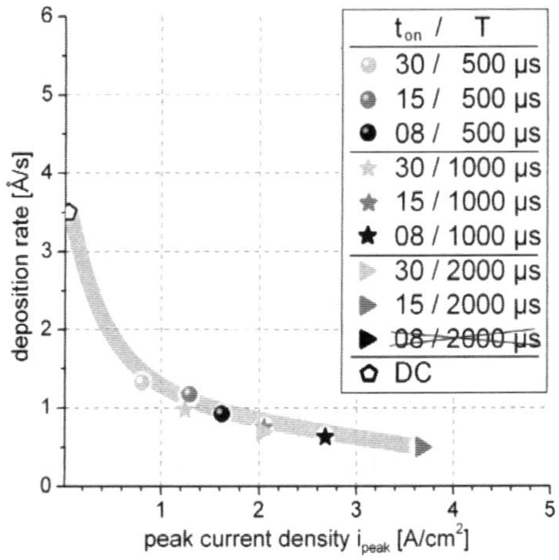

Figure 6. Deposition rate of Ti vs. i_{peak} for different pulse times at p_{Ar} = 1 Pa (data points with P_{av} < 250 W not included).

3.3. Ionization

By evaluation of t.a.OES data it was verified that an increasing peak current density i_{peak} basically produces an increase of the relative amount of single ionized Ti^+. Trends like this have been observed in many publications for several materials like e.g. in [2-6]. Figure 7 exhibits t.a.OES vs. i_{peak} results for several sets of pulse configurations and Ar-pressures. Furthermore this figure reveals, that the increase mentioned is more pronounced for higher Ar-pressure. Again it can be seen, that one and the same relative Ti^+ single ionization can be realized by several sets of parameters. Please note that in figure 7 data points of the same type do represent fixed pulse configurations at different Ar-pressures.

Rate and ionization data suggest, that at the same total energy input (P_{av} = 250 W constant) the more energy is fed into the ionization of the deposition particles (which comes along with higher i_{peak}), the less remains for the actual sputtering (lower deposition rate). This could (beside of the back attraction of ions to the target and other effects as described in several publications [2, 6-8]) be another reason for the rate loss observed with the use of HIPIMS. Christie stated that there is "ultimately a trade between target material ion fraction… and deposition rate" [7].

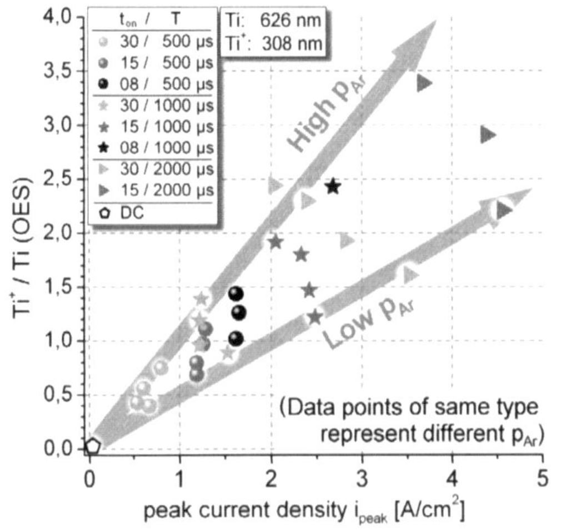

Figure 7. Relative single ionization of Ti vs. i_{peak} at different pulse times and Ar-pressures (data points with $P_{av} < 250$ W not included).

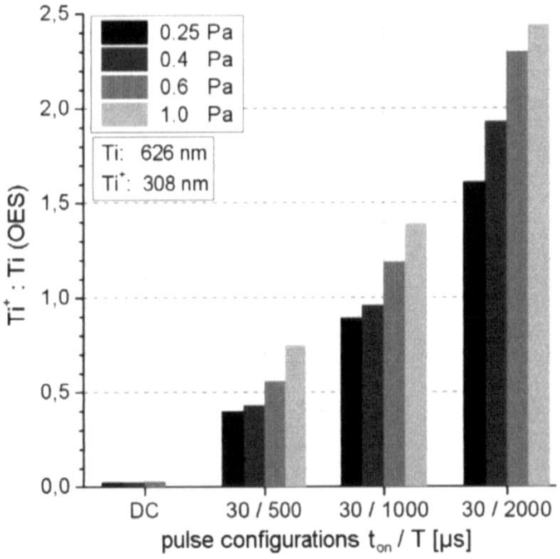

Figure 8. Relative single ionization of Ti at constant t_{on} and different T with comparison to DC-MS.

Due to the interconnection of the peak current density i_{peak} and the deposition rate shown in the previous section, a direct link between the increase of Ti-ionization and the decrease of the deposition rate can be stated, which is also supported by the publications already mentioned [2-6].

Any kind of decrease of the duty cycle (reducing t_{on} or extending T) was found to cause the relative Ti^+-ionization to increase. An example for this is shown in figure 8 with constant t_{on}= 30 µs and different T. A similar basic trend has been found and published for Cr [3]. Another information in this figure is, that an increase of p_{Ar} also gives rise to an increased relative Ti^+-ionization, independently of the pulse configuration. Since the target voltage U_T decreases with increasing p_{Ar} (and lower U_T was already observed to generally produce lower i_{peak} and thus actual lower ionization), a higher ionization efficiency as mentioned by [2] can be stated here. This in turn could presumably be attributed to a higher electron density and -impact rate. Additionally in figure 8 the great difference to the ionization in DC magnetron sputtering (DC-MS) is visualised. It can be seen that HIPIMS, even operated at short T and a duty cycle of 6 % (t_{on}/ T: 30/ 500 µs, i_{peak} = 0.54 – 0.8 A/cm^2), produces a far higher amount of single ionised Ti^+ than DC-MS. A transition point, as identified by Alami for Cr at i_{peak}= 0.57 A/cm^2 [3] (at which HIPIMS starts to produce reasonable amounts of ions at all) could not be observed here. If a transition point like this also exists for Ti, it should thus be found at i_{peak}-values far lower than 0.54 A/cm^2.

When the duty cycle is kept constant and t_{on} and T are varied in the same manner, a longer t_{on} causes a higher relative single ionization (figure 9, a similar trend was reported by [2]) and at lower p_{Ar} also a higher double ionization of Ti (figure 10). With increasing t_{on} basically a higher plasma density can evolve. An increasing p_{Ar} on the other hand causes a distribution of the available energy among more particles by a higher impact rate, which results in the double ionization to vanish more and more (an effect like this has also been found by [5] and [9]). At 1 Pa even no increase of double ionization can be detected for increasing t_{on} (figure 10).

As a consequence of the data shown a high single Ti^+-ionization can most effectively be combined with a low double ionization by using high Ar-pressure.

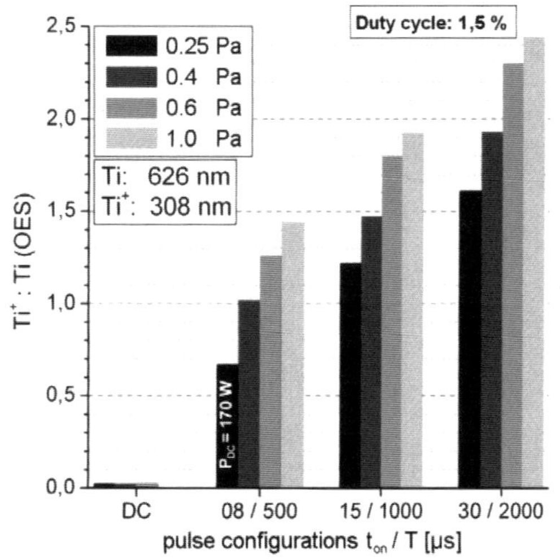

Figure 9. Relative single ionization of Ti at a constant duty cycle and different t_{on}.

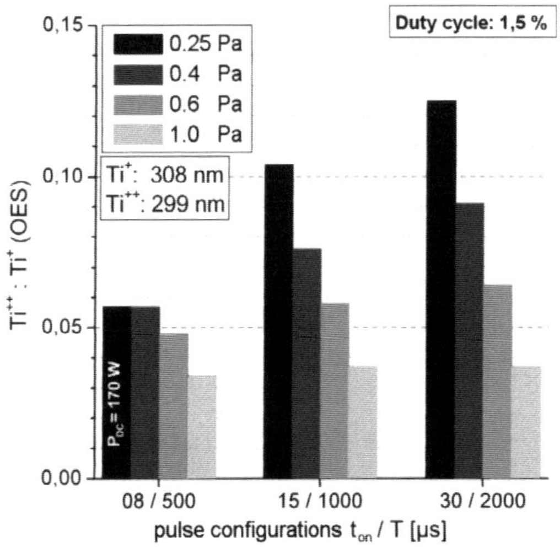

Figure 10. Relative double ionization of Ti at a constant duty cycle and different t_{on}

4. Conclusions

- In this work essential interrelations between pulse parameters, Ar-pressure and the properties of the HIPIMS sputtering process have been observed
- It could be shown that varying the pulse configuration (t_{on}, T) and the Ar-pressure in HIPIMS can lead to finding a viable compromise between ionization and deposition rate
- Even high duty cycles > 3 % were found to generate considerable ionization of Ti, along with low double ionization at high Argon pressure
- Deposition rate was found to decrease exponentially with increasing target peak current density at any constant Ar-pressure
- The energy fed into ionization of Ti and its higher mean kinetic energy [10] must be lacking for the sputtering itself, which could (beside back attraction of ions and other effects [2, 6-8]) be assumed as another reason for the decrease of the deposition rate observed in HIPIMS

This work was funded by the Bundesministerium für Bildung und Forschung (BMBF).

[1] Kouznetsov V et al. (1999) *Surface and Coatings Technology* **122** 2901

[2] Konstantinidis S et al. (2006) *Journal of Applied Physics* **99** 013307

[3] Alami J et al. (2006) *Applied Physics Letters* **89**, 154104

[4] Kudlacek P (2008) *Plasma Sources Science and Technology* **17** 025010

[5] Burcalova K et al (2008) *Journal of Physics D: Applied Physics* **41** 115306

[6] Ehiasarian A P et al. (2002) *Vacuum* **65** 147–154

[7] Christie D J (2005) *Journal of Vacuum Science and Technology* **A 23(2)** 330-335

[8] Emmerlich J et al. (2008) *Vacuum* **82** 867–870

[9] Hecimovic A, Ehiasarian A P (2009) *SVC Fall Bulletin* **28**

[10] Bohlmark J et al. (2006) *Thin Solid Films* **515** 1522-1526

HiPIMS "goes production", actual status & outlook

Oliver Lemmer, Werner Kölker, Stephan Bolz, Christoph Schiffers

CemeCon AG, D 52146 Wuerselen, Germany

Abstract. Hardness, toughness, and adhesion are today among the most important characteristics of PVD (Physical Vapor Deposition) coatings for machining tools that are used for highly demanding hard, high-speed, and dry machining. The continued development and improvement of materials constantly creates new and more demanding challenges for machining tools. In particular, materials that are especially hard or difficult to machine, materials that are prone to stress hardening, and materials with low heat conductivity require coatings with high thermal and oxidation stability in addition to the properties already mentioned. With a view to these properties, the new HiPIMS technology is currently being used to develop a new generation of coatings with significantly improved properties compared to conventional coatings. They are able to process hard-to-machine materials such as nickel-based alloys and stainless austenitic steels at lower costs - with comparatively significantly improved machining parameters and considerably less tool wear.

New impulses for machining

Recent years have seen an increased interest in pulsed sputter PVD processes among users. HiPIMS represents the latest development in the field of pulse technology. HiPIMS is a pulsed PVD method that achieves a high ionization of the sputtered target material.

The feeding of high-energy power pulses in the megawatt range into the target causes a plasma with a high charge carrier density (10^{19} m^{-3}) to form in front of it. This density is considerably higher than with conventional DC (direct current) sputter methods. The high ionization of the vaporized target material improves the coating structure and characteristics compared to conventional methods. The reason is that the energy and directionality of the positively charged particles that hit the substrate are favorably influenced by a negative voltage (bias voltage) that is applied to the substrate.

With the conventional DC sputter method, the ionization of the deposited target material is increased by raising the cathode power. However, this approach has its limits due to the increased thermal loads on the cathodes and substrates to be coated. HiPIMS solves this problem. The method employs very high-power pulses that only hit the target material very briefly. A relatively long subsequent "off time" ensures that the average cathode power remains low (1-10 kW), giving the target material time to cool down and providing high process stability.

Harder and tougher

Coatings coated with adjusted HiPIMS process parameters could have denser coating morphology and an increased ratio of hardness to Young's modulus compared to conventional PVD coatings. Whereas comparable conventional nano-structured (Ti,Al)N coatings have a hardness of 25 GPa and a Young's modulus of 460 GPa, the hardness of the new HiPIMS coating is greater than 30 GPa with a Young's modulus of 368 GPa. The ratio between hardness and Young's modulus is a measure of the toughness properties of the coating. The desirable condition is high hardness with a relatively small Young's modulus, such as can be found in HiPIMS coatings.

Corresponding author.
E-mail address: oliver.lemmer@cemecon.de (O Lemmer)

IOP Conf. Series: Materials Science and Engineering **39** (2012) 012003 doi:10.1088/1757-899X/39/1/012003

The comparatively higher thermal stability of the HiPIMS coating is due to the denser coating structure but also to the completely new material composition. In addition, the innovative coating method makes it possible to "anchor" the coating to the substrate and thus to double the adhesion of the coating compared to coatings used so far. This is particularly advantageous when sharp cutters are used for interrupted cuts, for example in maching of Superalloys.

HiPIMS in practical use

Within a very short development phase, CemeCon engineers have been able to transfer the HiPIMS technology to the CC800® system platform. The CC800®/HiPIMS with the new HiPIMS technology has been running in production for more than a year and has proven to be exceptionally stable and reliable. This has been made possible by the targeted development efforts of tool manufacturers and coating suppliers. In introducing HiPIMS technology into the machining tools industry, CemeCon is simultaneously creating a new generation of coating materials with extraordinary properties. Moreover, HiPIMS technology can be integrated into existing CemeCon coating units. This means these machines will be able to produce the latest coatings in the future.

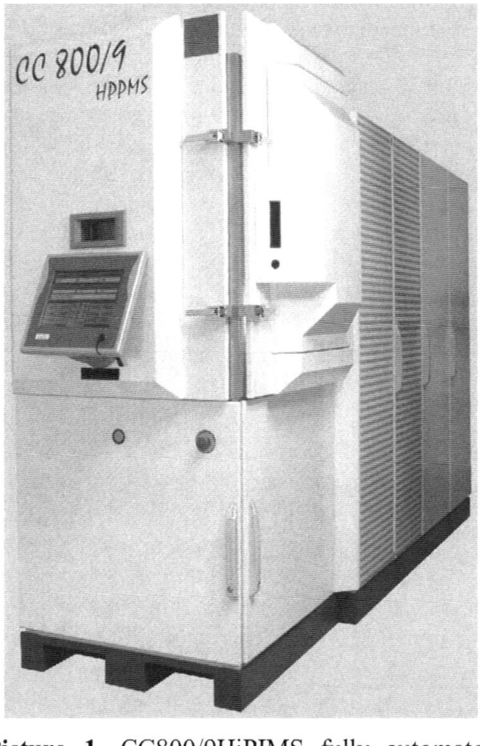

Picture 1. CC800/9HiPIMS fully automated coating equipment for production.

In reference to the high-energy power pulses used in the process, the new coating generation will be given the name "power nitrides". As with the super nitrides, the power nitrides family will include various new coating materials as well as enabling the production of tried-and-tested coatings (coating engineering). In the long run, HiPIMS has the potential to supersede existing PVD technologies.

New coating material

The first new coating material (working title HPN 1) was presented at the AMB 2010 and is available for test series on milling cutters, drills, and indexable inserts made from carbide. Extensive testing of the material, particularly with hard-to-machine materials such as stainless steel or Inconel, has been carried out and has yielded extremely good results (see differnet charts). The coating material also did very well in the machining of cast materials.

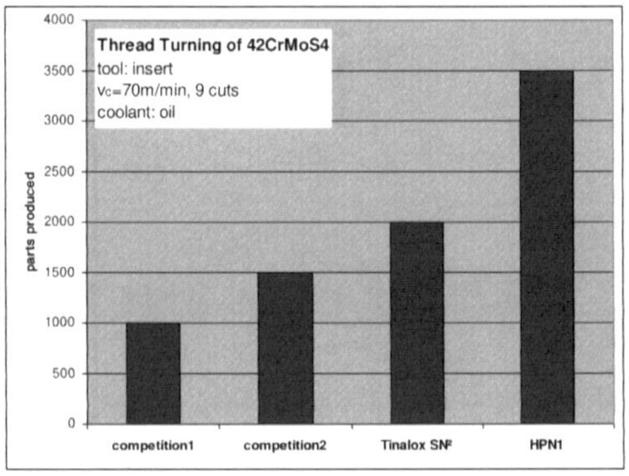

Chart 2. Thread turning of 42CrMoS4.

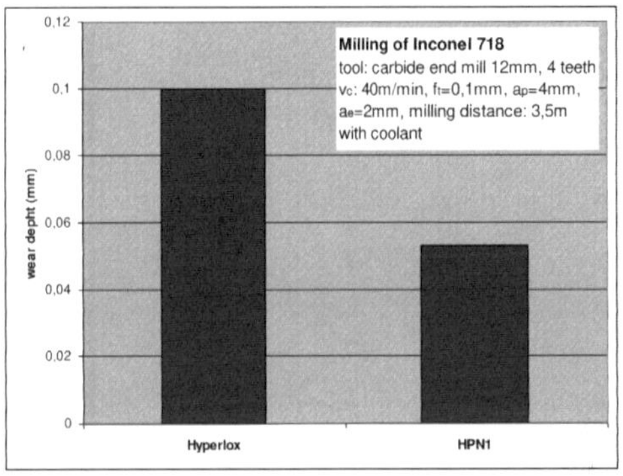

Chart 3. Milling of Inconel.

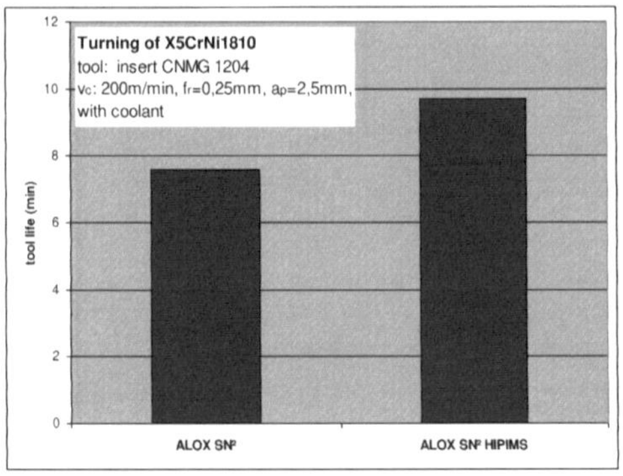

Chart 4. Turning of X5CrNi1810.

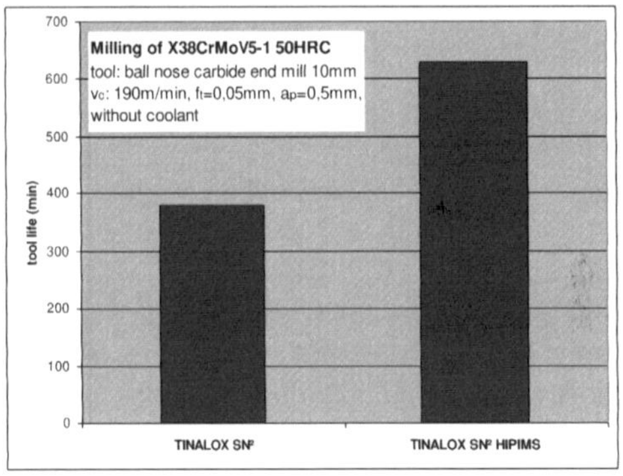

Chart 5. Milling of X38CrMoV5-1.

Conclusion and outlook

To be able to continue to offer their users potential for improvement in the future, CemeCon took the next step in coating technology and validated HiPIMS: Optimized adhesion, denser coating morphology, and a lower Young's modulus with, at the same time, increased hardness – these are just some of the highly promising possibilities of HiPIMS.

Keywords:

- hard or difficult to machine materials
- high-energy power pulses
- coating engineering
- power nitrides

IOP Conf. Series: Materials Science and Engineering **39** (2012) 012004 doi:10.1088/1757-899X/39/1/012004

Domino Platform: PVD Coaters for Arc Evaporation and High Current Pulsed Magnetron Sputtering

J. Vetter, J. Müller, G. Erkens

Sulzer Metaplas, Am Böttcherberg 30-38, 51429 Bergisch Gladbach, Germany

Abstract. AlTiN and CrN coatings were deposited in hybrid DOMINO platforms by magnetron sputtering (DC-MS, DC-MS+HCP-MS, HCP-MS) and vacuum arc evaporation. The ion cleaning was done by the AEGD process. The coating rates and the energy efficiency of both deposition processes were compared. The roughness effects of the different coating types were discussed. Preliminary results of the change of pulse characteristics during simultaneously running of HCP-MS plus vacuum arc evaporation are shown.

Keyword: Arc, sputtering; HCP-MS, energy, roughness, hybrid

1. Introduction

The vacuum arc evaporation and the magnetron sputtering are the PVD deposition methods of Sulzer Metaplas to produce tribological and other functional coatings. The application of vacuum arc evaporation is the dominating PVD method for tool coatings and high performance component coatings. Besides advanced hard coatings also upgraded arc evaporators like APA (Advanced Plasma Assisted) are available. Classical DC-magnetron sputtering (DC-MS) is mainly used for a-C:H:Me coatings or for the deposition of interlayers for a-C.H:X coatings. Although the DC-magnetron sputtering process often results in smoother coatings the number of coating systems in industrial use is much lower than that of the arc evaporation. The reasons for that are the advantages (stability, productivity, costs, and coating properties) of the arc evaporation process in industrial scale.

Brief description of the arc evaporation deposition process

The vacuum arc evaporation is a self-sustaining discharge under vacuum condition. The arc process starts with the striking of a high current, low voltage arc on the surface of a cathode. The current density within a cathode spot (diameter ca. 10 µm) goes up to 10^7 A/cm^2 assuming a minimal spot current of 10 A [1]. The temperature at the cathode spot is extremely high resulting in a jet of vaporized cathode material. The localized cathode spot exists only for a short period of time. The spots jump to another area. The plasma jet consists of multiple charged ions, neutral atoms and macro-particles (droplets). A high level of ionization depending on the cathode material (30-100%) is observed [3]. Reactive gases are used to deposit nitrides, carbonitrides, oxynitrides and oxides. The reactivity is also given at low temperature due to excitation, dissociation and ionization. To eliminate the emitted droplets different filter systems are used [3].

Corresponding author.
E-mail address: joerg.vetter@sulzer.com (J Vetter)

Brief description of the magnetron sputtering deposition process
The DC magnetron sputtering (DC-MS) is based on the atomization of the target material by an energetic bombardment of its solid surface by ions. The sputtering process is possible in either under vacuum condition, or in inert gas or in reactive (inert plus reactive) gas atmospheres. The electrical conductivity of the target material has to be sufficient high. The ionization is mostly dominated by the sputter gas ionization (mostly Ar). Two discharges are of special interest:

- A) The triggered self-sustained magnetron sputtering without a gas burning is the plasma of ionized sputtered atoms [4].
- B) The high current pulsed magnetron sputtering HCP-MS burning in a discharge consisting of a mixture of ionized argon with a high ionization of the sputtered atoms [5], [6], [7]. This higher ionization level of the magnetron sputtering process opens up new possibilities to tailor coating properties. The HCP-MS processes have different trade names: HIPIMS, HIPIMS+, HPPMS, MPP, HIPAC and others.

A demand of R&D and small volume production systems to run HCP-MS processes is obvious. The solution provided by Sulzer Metaplas is the DOMINO platform (mini, S, L type) configured with both HCP-MS and APA arc evaporators as hybrid systems. Preliminary results of both processes as well as of the hybrid process will be presented.

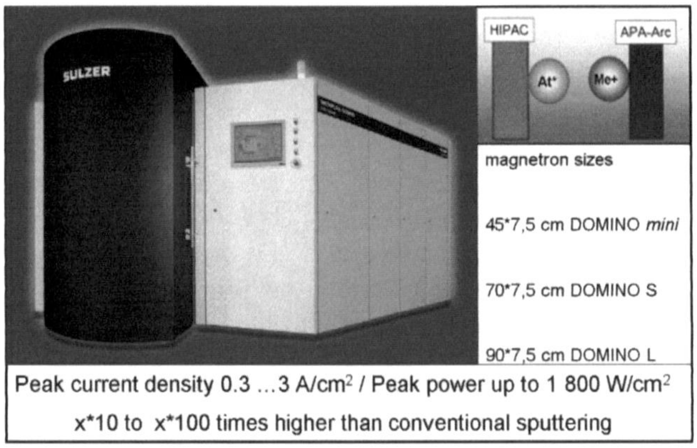

Figure 1. METAPLAS-DOMINO platforms for the hybrid set up: HCP-MS and vacuum arc evaporation with different chamber and magnetron sizes: HIPAC (High Ionized Plasma Assisted Coating: a trade name of Sulzer Metaplas for HCP-MS).

2. Experimental set up

Two important industrial coating types were deposited: AlTiN coatings using a target composition developed by Sulzer Metaplas (Al66at%Ti34at%) [8] and CrN coatings using pure Cr targets. Two different deposition methods were applied: magnetron sputtering and arc evaporation. Magnetron sputtering of the AlTiN coatings was made in the DOMINO mini using two magnetrons mounted in a closed field configuration. The magnetron with an unbalanced magnetic field set up had a size of 7.5 cm in width and 450 mm in length. The used pulse unit allowed a maximum voltage of 1000V/1000A. The magnetron discharge was running in 3 different modes: two magnetrons in the DC-mode, one magnetron in DC mode and the other in HCP-mode (HIPAC), both magnetrons in the HCP-mode. The distance between the substrates and the target surface was 60 mm. The bias voltage was set to 50 V, the total pressure was about 1 Pa, Ar:N_2 flow of 120:120 sccm. The average power applied to each target was constant with 7 KW (20 W/cm^2). Fig. 2 illustrates the different pulse modes.

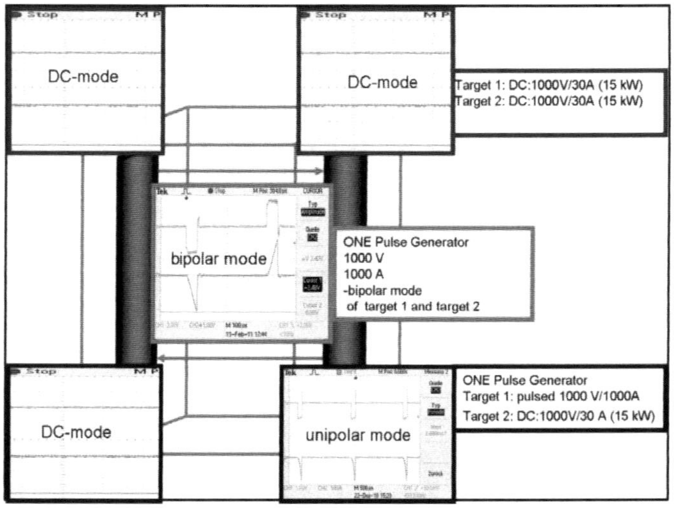

Figure 2. Possible modes of magnetron sputtering in the DOMINO mini.

To compare the growth rate with the growth rate of the arc evaporation also AlTiN-coatings were deposited by industrial standard parameter. The distance to the substrate holder was 140 mm. CrN coatings were deposited by HCP-MS and vacuum arc to compare the adhesion level and roughness of the coatings in a DOMINO S system. The deposition parameters were the following: total pressure 0,9 Pa, Ar:N_2 flow of 250:80 sccm, bias 50V, pulse parameters 150 µs on 1500 µs off (f= 606 Hz). The CrN arc coatings were deposited under industrial standard parameters. The ion cleaning was carried out by AEGD (Arc Enhanced Glow Discharge) before HCP-MS and arc evaporation [9]. Preliminary investigations of pulse shape modifications during the hybrid process combining HCP-MS and vacuum arc evaporation using a silicon sputtering target and Cr arc cathodes were done by oscilloscope measurements in the DOMINO S using targets of 7.5 cm in width and 70 cm in length.

3. Results

3.1. Deposition rate of AlTiN coatings in HCP-mode

Fig. 3 shows the deposition rates achieved by magnetron sputtering in the different operation modes. The highest deposition rate was measured for DC-sputtering (14 A, 500V) with a current density of 0.04 A/cm^2. A decrease was observed when operating in the mixed mode; one target with DC-MS and the other by HCP-MS. A sharp drop was measured when a high peak power was applied at both magnetrons running in the HCP-mode. The pulse parameters for the high ionized plasma operation were 80 µs on, 500 µs off (f=1724 Hz). The maximum peak current with 600 A results in a current density of 1.8 A/cm^2 and the maximum peak power of about 1 170 W/cm^2.

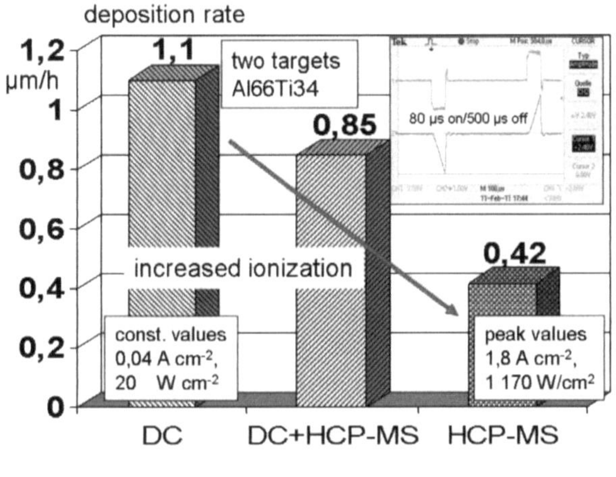

Figure 3. Deposition rate of AlTiN in different sputtering modes (3-fold rotation).

3.2. Energy consumption and deposition rate

A standard industrial (Al66at%Ti34at%)N coating was deposited by two APA evaporators in order to compare the energy consumption to get a specific deposition rate of the different sputtering modes with the energy consumption of the arc evaporation. The same substrate holder was used (3-fold rotation). The deposition rate was measured over a height of about 310 mm (magnetron: ca. 70% of the target length, arc: ca. rim of the arc evaporators). The measured deposition rate [nm/h] was divided by the total energy consumption of the particle sources [KW]. The results are shown in Fig. 4.

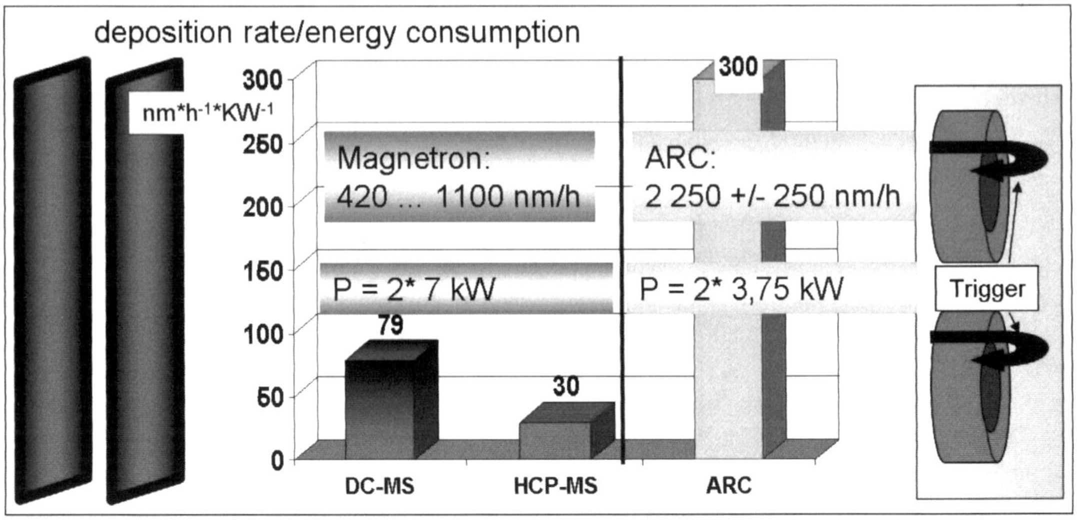

Figure 4. Deposition rate divided per energy consumption of different sputtering modes and arc evaporation.

Obviously the energy consumption to get the same deposition rate increases if the sputtering is switched from the DC- mode to the high ionized HCP-mode. The energy consumption to get the same deposition rate increases by a factor of 2.5. The outstanding low energy consumption of the arc is significant. The required energy to get the same deposition rate like the HCP-MS is up to 10 times lower than for the HCP-MS.

3.3. Roughness

The adhesion and the roughness for CrN coatings deposited by HCP-MS and arc evaporation are shown in Fig. 5. Both coating methods show the same high adhesion achieved after AEGD-ion cleaning. The roughness is typically much lower for the HCP-MS coating than for the arc coating. The roughness of the arc coating decreases significantly when applying an industrial standard post treatment (brushing). However it is still slightly higher than that of the HCP-MS coating.

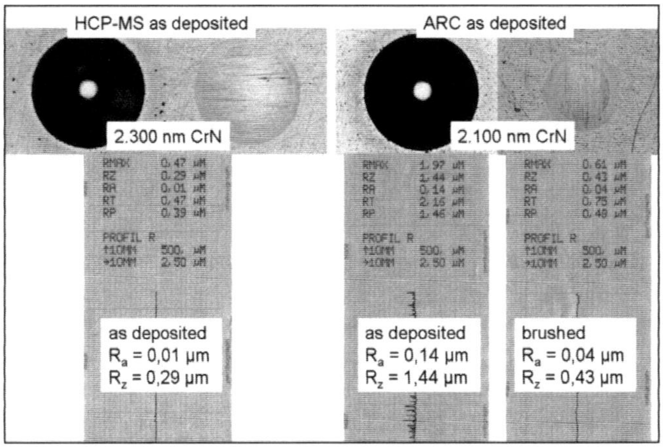

Figure 5. Roughness of as deposited CrN coating: HCP-MS in as deposited state, arc in as deposited and post treated state.

3.4. Delay of pulse current

Fig. 6 shows two pulse forms for a Si-target (f=500 Hz, 150 µs on) in the reactive mode which have significantly different delay times of the current rise. If the silicon target is running without the arc evaporation than the delay time is of about 25 µs. If the arc evaporators are switched on, than that delay time is nearly zero. Strong plasma activation is always present in the chamber resulting in a fast current increase.

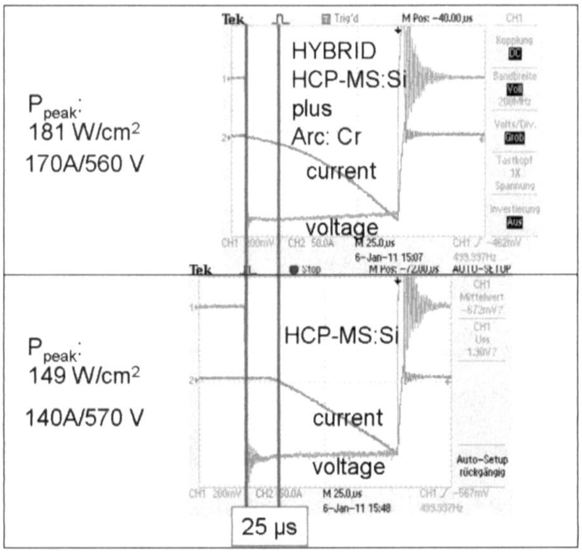

Figure 6. Delay time of the current in reactive HCP-MS of SiNx process and in a reactive hybrid process by HCP-MS of Si and Cr by arc evaporation.

4. Discussion

Table 1 shows selected general process data's and industrial relevant data's for DC- MS, HCP-MS and arc evaporation. In addition to the well known data's the energy consumption is added. It has to be pointed out that this comparison is made under the view of industrial applications of hard coatings for wear and friction reduction.

4.1. What are the advantages of HCP-MS compared to Arc for wear/friction reduction coatings

- A) It has been shown that HCP-MS is able to produce well adherent and smooth coatings. That means for micro tools and components it is an alternative to the filtered vacuum arc evaporation. It should be mentioned, that the low roughness requires both an optimization of process parameters and a clean surrounding inside and outside the chamber, otherwise growth defects will be generated by dust particles [10].
- B) The target material variety is larger than for the arc evaporation. Thus HCP-MS is the solution for target materials which are difficult or not evaporable by the vacuum arc evaporation, e.g. TiB2, SiC, B4C.

4.2. What are the disadvantages of HCP-MS compared to Arc for wear/friction reduction coatings?

- A) It has been shown that the deposition rate is significantly lower than that of the arc evaporation. The shown decrease of the deposition rate of HCP-MS in comparison to DC-MS is caused by the high power densities applied in the described experiment. The process parameters shall be tailored to apply an optimum of power densities for the required coating properties and by adjusting the reactive gas process to maximize the deposition rates.
- B) In addition, the energy consumption to get the same deposition rate is up to a factor 10 higher than for the arc evaporation. This tendency was also shown for Cu coating processes [2].

4.3. Which potential does hybrid systems have?

The DOMINO platform has both the possibility to deposit coatings either by pure HCP-MS and its combination with DC-MS or by arc evaporation to find the optimal coating process for different applications. It should be mentioned, that the droplet content and the resulting roughness of an arc deposited coating is acceptable for a lot of applications without any post treatment. In several applications like for forming tools and plastic moulds a post treatment has to be applied. One not in detail investigated potential of HCP-MS is the hybrid process with vacuum arc evaporation. The possibility to use target materials which are difficult or not evaporable by arc opens the window to deposit dedicated nanocomposites by doping or multilayers (also in the nanoscale) as well as top layers. First steps were done by Sulzer Metaplas.

Table 1. Basic comparison of arc and magnetron deposition.

	Target material flexibility	Ioni-zation	Rate	Energy efficiency	Main application wear/friction	* Classical hard coatings	*DLC with PVD process
DC-MS	++++	+ max 10%	+++	++	DLC parts decoration (tools)	5%	99 %
HCP-MS	+++	++ may exceed 50% [7]	++	+	not defined smooth and dense coatings	?	?
DC-Arc	++ limited arc materials	+++ up to 100 % [2,3]	++++	++++	tools, components (decoration)	95%	1%
Arc Filter	++ mostly carbon	++++ 100%	+	+	Components (tools)	less 1%	less 1%

*percentage of application is estimated from market observation, not based on statistical evaluation

5. Summary and conclusions

- 1. The DOMINO platform gives the possibility to investigate specific process parameters and coating properties for DC-MS, HCP-MS in comparison to the arc evaporation.
- 2. The DOMINO platform allows to run HCP-MS and vacuum arc evaporation as a hybrid process.
- 3. The deposition rate of HCP-MS is lower than that of DC-MS and significantly lower than that of the vacuum arc in the investigated configuration.
- 4. The energy consumption to get a specific deposition rate is for HCP-MS higher than for DC-MS and much higher than for the vacuum arc evaporation (up to factor 10).
- 5. Preliminary investigations of the pulse shape during simultaneous operation of arc evaporation and HCP-MS show a significant change in the pulse form.

References

[1] Siemroth P., Schultrich B., Schülke T., Surf. Coat. Techn. 74-75 (1995)92-96
[2] Rother B., Vetter J., Wiss. Z. d.TU Karl-Marx-Stadt 31(1989) H3, p. 433-445
[3] Boxman R.L., Martin P.L., Sanders D.M.: Handbook of vacuum arc science and technology, Noyes Publications, New Jersey, 1995
[4] Dorodnov A.M, Petrosov B.A, Sov. Phys. Tech. Phys 26(3)1981, 304
[5] Khodachenko G.V., Mozgrin D.V., Fetisov I.K. patent RU2058429, 1993
[6] Helmersson U., Lattemann M., Bohlmark J., Ehiasarian A.P., Gudmundsson J. P., Thin Solid Flms, 513(2006) 1-24
[7] Anders A., Surf. Coat. Techn. 204(2010) 2864-2868
[8] Vetter, J., Journal of Advanced Materials. Band 31, Nr. 2, 1999, S. 41–47
[9] Vetter J., Perry A.J., Surf. Coat.Techn. 61(1993)305-309
[10] Vetter J., Stüber M., Ulrich S., Surf. Coat. Techn. 168(2003) 169-178

IOP Conf. Series: Materials Science and Engineering **39** (2012) 012005 doi:10.1088/1757-899X/39/1/012005

Exotic Highly Ionized Pulse Plasma Processes

R Bandorf, G Bräuer
Fraunhofer Institute for Surface Engineering and Thin Films, Bienroder Weg 54 E, Braunschweig, Germany

Abstract. Highly ionized pulse plasma processes (HIPP processes), like high power impulse magnetron sputtering (HIPIMS) also known as high power pulse magnetron sputtering (HPPMS), modulated pulse power sputtering (MPP), as well as further modifications involving increased ionization in sputtering have been investigated and matured within the last decade. On European level a concerted action focusing the efforts and bringing together the leading experts was set up and is actually promoting the transition of HIPP processes to industry. Besides the well know approaches of HIPIMS/HPPMS and MPP further modifications for ionized PVD are under development. This paper focuses on HIPP processes and new technological approaches besides the well-known "big players". New approaches combining technologies like hollow cathodes in different modifications, coils for inductive coupling, arc sources for "gasless sputtering" or a modified process for constant-current HIPIMS will be presented.

1. Introduction

High power impulse magnetron sputtering (HIPIMS) or also known as high pulse power magnetron sputtering (HPPMS) and modulated pulse power sputtering (MPP) are recent technologies in the field of physical vapour deposition (PVD) to realize higher plasma densities for superior or new coatings [1]. Since the mid 90's the development focusses on pulsed processes with pulse duration of 1 µs to several 100 ms and frequencies of a few ten to 100 Hz [2-5]. Due to the low duty cycle the peak power and current are significantly increased by orders of magnitude while the average power and current remain similar to conventional sputtering processes. Furthermore the obtained plasma and electron density is significantly higher and therefore the resulting ionization of the sputtered species [2-5]. From the academic research the technology has moved to industrial applications [6-8]. Besides the industrialization novel approaches combining new pulse power generators and deposition sources arise. Therefore this paper will give an overview of new technological developments in the field of HIPP processes besides established HIPIMS/HPPMS and MPP.

For the realization of new HIPP approaches, both available HIPIMS power supplies and tailor made solutions are used. Apart from conventional magnetron sputtering cathodes the emphasis is on sputtering diodes without magnetic field, hollow cathode arrangements as used for gas flow sputtering or plasma jet, coupled discharges using an electrically pulsed coil or a triggered arc source, as well as a new generator concept.

Corresponding author.
E-mail address: ralf.bandorf@ist.fraunhofer.de (R Bandorf)

2. New HIPP technologies

The common factor of all the new approaches is to overcome drawbacks of existing HIPP processes by improving the deposition rate, removing constraints by the magnetic field, avoiding the process gas and realizing a self-sustained discharge or tailoring the ion to neutral ratio by avoiding long transition states in the pulse. The presented technological approaches will be the combination of HIPIMS power supplies with hollow cathode arrangements (HIPP-GFS, plasma jet, HPPS), inductively coupled impulse sputtering (ICIS) using an rf-coil operating with bursts, a triggered arc source acting as kick-starter for gassless sputtering, and a new generator concept for long constant current regime magnetron sputtering.

2.1 Hollow Cathode Processes in HIPP-mode

Using pulse power supplies to operate different hollow cathode processes results of improved plasma and coating properties are reported for plasma jet, gas flow sputtering (HIPP-GFS), and magnetic hollow cathode arrangement, called high-power pulsed sputtering (HPPS) glow plasma.

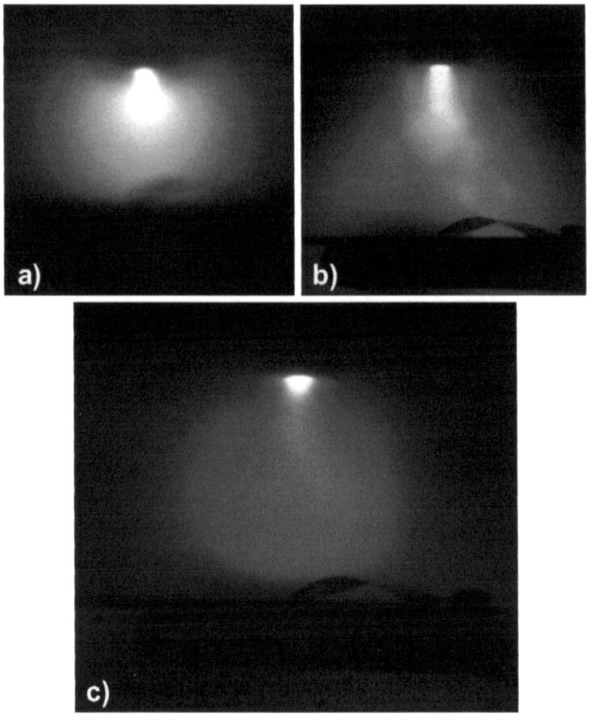

Figure 1. Plasma jet operated in a) DC, b) HIPIMS, and c) combined DC and HIPIMS mode

Starting with the plasma jets it is known from literature, that RF driven plasma jets can be used both, for reactive and non-reactive sputtering [9,10]. By choosing the right dimensions of the plasma jet, i.e. a tubular hollow cathode, this system offers high potential for inner coating of tubes. Nevertheless generating a stable discharge and at the same time limiting the heat generation is a challenging task. To solve this problem a combination of a HIPIMS power supply and a plasma jet was investigated in the frame of a COST short term scientific mission from J Oksanen at Linköping university. For copper, titanium, and carbon investigations were carried out applying DC, HIPP, and combined DC and HIPP to the plasma jet. Figure 1 shows pictures of the plasma jet working with Ti in a) DC, b) HIPP, and c) DC + HIPIMS mode. From the brightness and the distributuion of the plasma a first estimate was made for the quality of the process. The DC process was not satisfying. The pure HIPP operation resulted in strange plasma distribution that should be further investigated. The combination of DC and HIPP resulted in a uniform glow, while the colour indicated that the plasma is dominated by the metal species. Further results are expected to be published by J Oksanen et al.

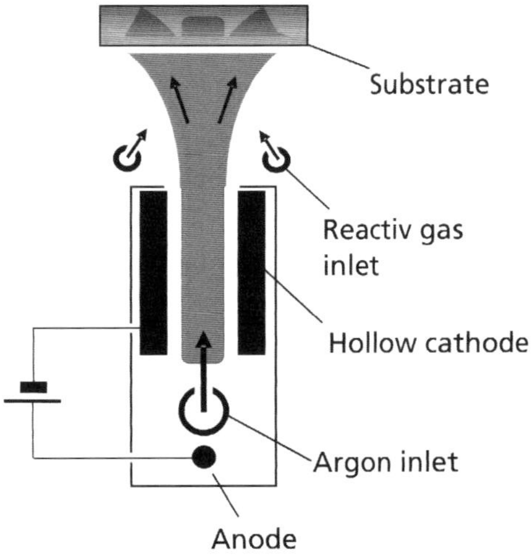

Figure 2. Schematic of Gas-flow sputtering

Figure 3. GFS source in operation

The second modification of hollow cathode processes is gas flow sputtering. Gas flow sputtering GFS is a technology using a hollow cathode arrangement of either a tubular target or two planar cathodes facing each other and a laminar gas flow through the source. Thus GFS offers a PVD technology operating in coarse vacuum with deposition rates of several ten micron to more than 100 µm per hour depending on the material [11,12]. Since the processes are running at typical pressures of 0.5mbar the argon flux is led through the source preventing target poisoning and realizing convectional material transport. Mostly the coatings are columnar and due to the high working pressure increased ionization can improve the densification of deposited films.

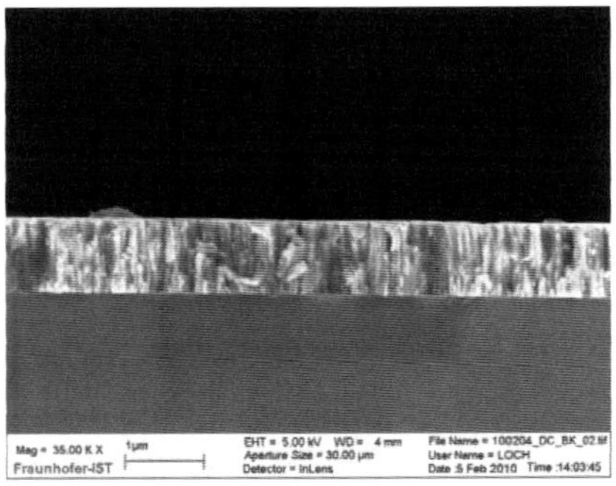

Figure 4. SEM cross-section of a 2 kW DC sputtered NiFe film

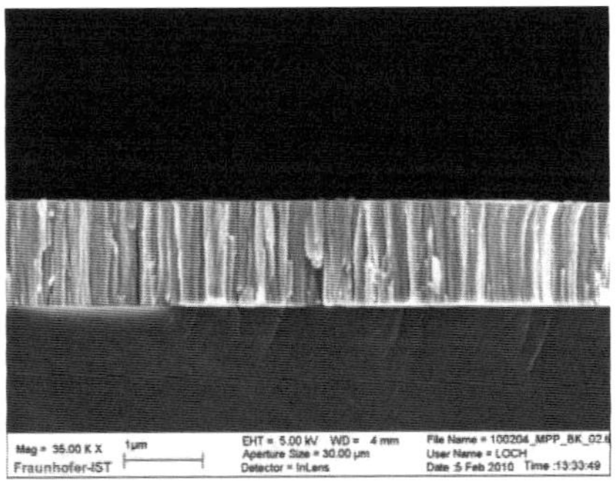

Figure 5. SEM cross-section of a 2 kW HIPP sputtered NiFe film

To realize HIPP-GFS the GFS-source was connected to a commercial HIPP power supply (Hüttinger, Melec, ZPulser) and operated in HIPP mode [13]. Preliminary tests with chromium showed that comparing DC and HIPP mode, in the latter case ions were detected by optical emission spectroscopy while in the former case only neutrals contributed to the film growth. Using the same gas flow, deposition time, and average power it was found for NiFe films that the HIPP-GFS films were

deposited at a 30% increased rate compared to the DC coatings. Furthermore the film structure, as shown in figure 4 and 5 was altered to a more pronounced columnar growth with fine grain structure on the columns in case of HIPP-GFS [13].
As third approach using a hollow cathode configuration so called High-Power Pulsed Sputtering (HPPS) Glow Plasma was investigated by K Azuma et al.. The group presented a sputtering source for arc free high power pulsed sputtering (HPPS) [14]. The basic principle is a hollow cathode arrangement of two targets on the same potential facing each other. The plasma is generated using a Penning discharge. An electric field and a magnetic field are parallel to each other. The presented plasma source has a small size of 60×60×60 mm^3 and can realize a power density in the order of a few kW/cm^2 over a gap length of 10 mm between both targets. The current densities of more than 2 A/cm^2 are reported. The estimated plasma density is in the order of 10^{13} cm^{-3} [14].

Comparison of the HPPS mode and the hollow cathode discharge mode (HCD), realized by removing the magnets behind the targets showed that the HPPS mode generated more ions, especially metal ions, while the HCD mode showed no significant optical emission lines from any species (figure 6) [15]. Due to instabilities in the HCD mode the acquisition of OES data was problematic and no reliable information was gained in contrast to the HPPS mode. The HPPS plasma has advantages in power density and the production of metallic ions but the plasma is confined to the gap between the targets due to magnetic trapping. Compared to HPPS the current density of HCD mode is only 10%, i.e. less than 0.1 A/cm^2 compared to 0.8 A/cm^2.

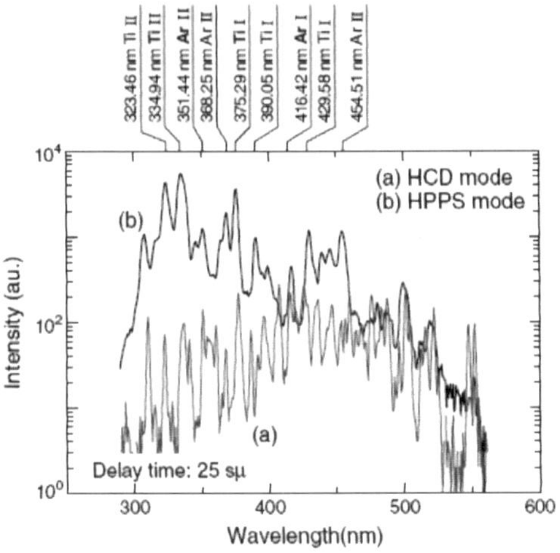

Figure 6. OES spectrum of HCD and HPPS mode [15] [1]

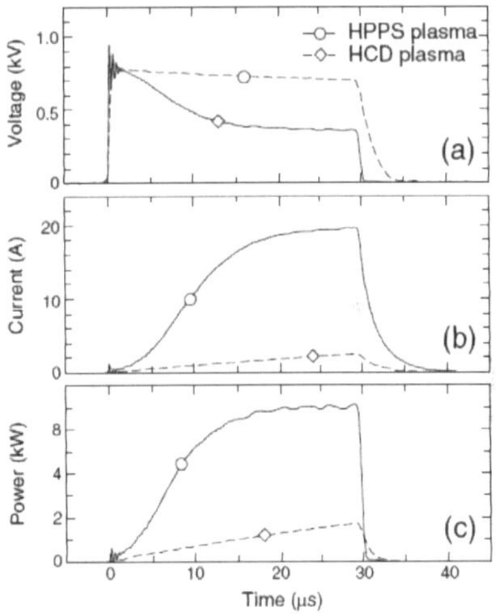

Figure 7. Voltage, Current, and Power evolution during single pulse in HPPS and HCD operation [15] [1]

2.2 Inductively Coupled Impulse Sputtering (ICIS)

K Yukimura et al. describe the realisation of highly ionised plasma without using a magnetic field. For this approach a pair of electrodes with applied pulsed DC is additionally superimposed by RF plasma that is generated by an external coil [16]. The coil is using a 200μs pulse at 200kHz-RF and 3kV applied voltage. Depending on the delay time between the RF-pulse and the DC-pulse a current up to

[1] Reprinted from Surface and Coatings Technology, Vol. 206, Iss. 5, K. Azuma, R. Mieda, K. Yukimura, H. Tamagaki, T. Okimoto, Comparative study of high-power pulsed sputtering (HPPS) glow plasma technique using Penning discharge and hollow-cathode discharge, 938-942, 2011, with permission from Elsevier.

35 A, corresponding to 1.6 A/cm^2, is measured. Figure 8 show the gap current and voltage of a DC-pulse following with 101 μs delay after the ignition of the RF-pulse [16].

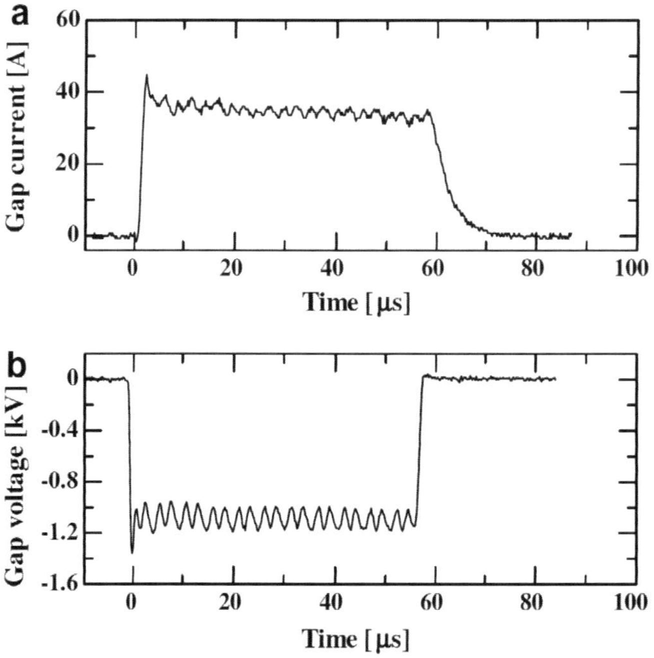

Figure 8. Gap current (a) and voltage (b) at a delay time of 101 μs between RF-pulse and DC-pulse [16] [2]

Further development of the technology is reported by D Loch et al moving the RF-coil into the process chamber and placing it in front of a cylindrical cathode. The plasma is ignited by the RF-pulse. Results are reported elsewhere [17].

2.3 Gasless sputtering

It was shown that the discharge in HIPIMS can switch from initial phase, dominated by gas ions to a later phase dominated by the target ions [18-21]. Furthermore the operation of the sputtering target in the absence of the sputter gas promises ultraclean coatings and other applications [20].Utilizing a fraction of the target ions that are back attracted to the target for further sputtering offers the chance to reach operation of the system operate in sustained self-sputtering.

---

[2] Reprinted from Nuclear Instruments and Methods in Physics Research B, Vol. 267, K. Yukimura, A. P. Ehiasarian, Generation of RF plasma assisted high power pulsed sputtering glow discharge without using a magntic field, 1701-1704, 2009, with permission from Elsevier.

IOP Conf. Series: Materials Science and Engineering **39** (2012) 012005 doi:10.1088/1757-899X/39/1/012005

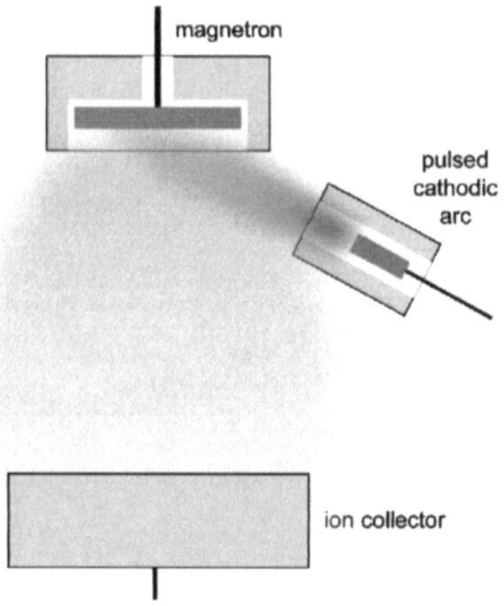

Figure 9. Schematic of the gasless sputtering; pulsed arc source was used as kick-starter for the magnetron discharge [20] [3]

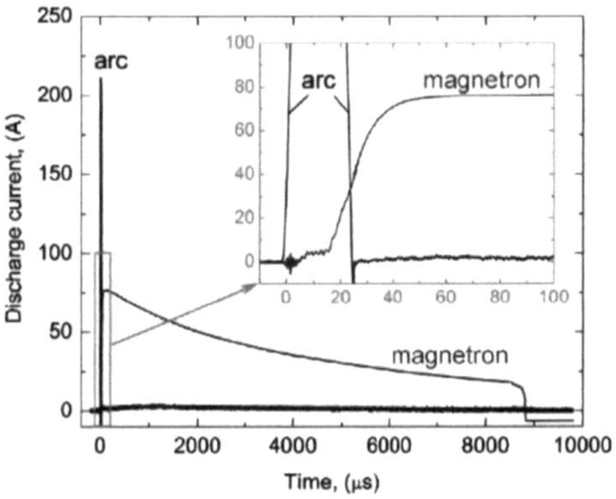

Figure 10. Discharge current of gasless sputtering; first ~ 20 μs high current pulse by arc source followed by self-sustained magnetron discharge up to 9 ms [20] [3]

Andersson and Anders showed for copper that HIPIMS processes can be realized in high vacuum, i.e. without additional inert gas (Figure 9 and 10). For the ignition a short pulse by a cathodic arc source was needed before the magnetron was operating in pure self-sputtering. Very high plasma densities of about $6x10^{18}$ m^{-3} were measured [20].

2.4 Constant Current HIPIMS

The proof of the high ionisation in HIPIMS is often referred to optical emission spectroscopy data and the visibility of ion emission lines [22,23]. Investigation of the emission lines in pulse time mode showed that the ion lines of titanium followed the current evolution [24]. Since the ionization degree is in transition during the pulse the question arises how to tailor the discharge to get a constant ion to neutral ratio, since this ratio determines the resulting film properties [25,26]. Furthermore it was shown by G Greczynski et al that the peak target current density and correlated with it the ion-to-neutral ratio influence the film growth [27]. Since in HIPIMS mostly no steady state of the current is observed during the pulse the ion-to-neutral ratio is also continually changing, as also seen by earlier OES measurements [24].

To keep the ion-to-neutral ratio nearly constant a setup with different voltage driven sources, i.e. two different capacitor banks were used to create a fast current rise in the beginning of the pulse (up to a few ten microseconds) and then maintain the current constant for the remaining pulse duration (up to a few hundred microseconds).

[3] Reprintet with permission from J. Andersson, A. Anders, Gasless sputtering: Opportunities for ultraclean metallization, coatings in space, and propulsion, Appl. Phys. Lett. 92 221503, 2008. Copyright 2008, American Insitute of Physics.

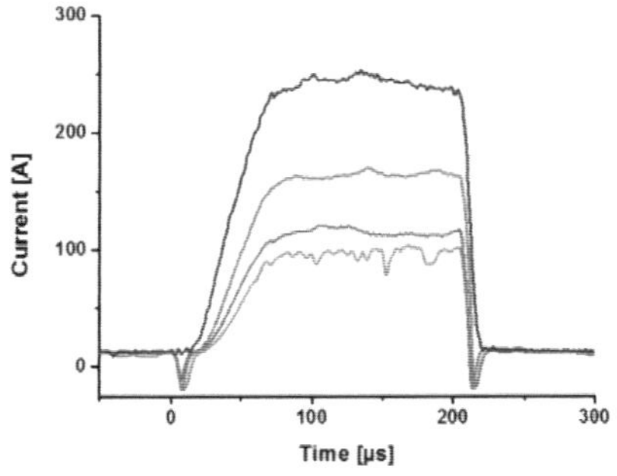

Figure 11. Different current levels (90 A, 110 A, 160 A, 240 A) realized by matching of two voltage sources

By tailoring the voltage levels for the long steady state phase and the additional voltage for fast current rise a rectangular current shape during the pulse can be realized. The modulation of the current with respect to a desired current plateau can be adjusted within a window of a few amps scattering (figure 11). Further details will be reported elsewhere [28].

3. Summary

Highly ionized pulse plasma processes, especially HIPIMS/HPPMS and MPP are already used in industrial scale. Additionally a variety of new approaches utilizing HIPP processes by combining existing pulse generators with technologies other than magnetron sputtering are proposed and discussed. Regarding high deposition rate and directional coating the combination of HIPP power supplies and hollow cathode arrangements (HPPS, HIPP-GFS, HIPIMS + plasma jet) were evaluated by different groups. Improvements of the resulting films are reported and further progress is expected. Also the combination of excitation by a pulsed coil and a diode seems promising for realization of ionized plasmas. The gasless sputtering, taking advantage of the sustained self-sputtering broadens the window for future applications for the synthesis of highly pure metals avoiding any side effects by any process gas. Finally a new approach for pulse mode at the cathode of a magnetron was presented realizing a pulse shape of the current with a short initial rise time and keeping a plateau level for the majority of the pulse duration. Corresponding to OES data a constant ion-to-neutral ratio is expected for a constant current level. The future will show the impact new approaches of HIPP technology can realize.

4. Acknowledgement

The authors acknowledge the financial support and knowledge exchange provided within the frame of COST Action MP0804 “Highly Ionised Pulse Plasma Processes – HIPP Processes” for hosting and attending meetings. Furthermore the authors like to thank J Oksanen for the permission to use data from his short term scientific mission report within COST Action MP0804 (STSM-7145 “The application of high-power pulses on a hollow cathode sputter discharge”). Finally the authors acknowledge the reprint permission referenced in the paper by Elsevier and the American Institute of Physics.

5. References

[1] Sarakinos K, Alami J and Konstantinidis S 2010 *Surf. Coat. Technol.* **204** 1661
[2] Mozgrin D V, Fetisov I K and Khodachenko G V 1995 *Plasma Phys. Rep.* **21** 401
[3] Fetisov I K, Filippov A A, Khodachenko G V, Mozgrin D V and Pisarev A A 1999 Vacuum 53 133
[4] Bugaev S P, Koval N N, Sochugov N S and Zakharov A N 1996 XVIIth International on Discharges and Electrical Insulation in Vacuum 1074
[5] Kouznetsov V, Macak K, Schneider J M, Helmersson U and Petrov I 1999 Surf. Coat. Technol. 122 290
[6] Bobzin K, Bagcivan N, Immich P, Bolz S, Fuß H G, Cremer R 2009 Plasma Process. Polym. 6 S124
[7] Münz W D, Schenkel M, Kunkel S, Paulitsch J, Bewilogua K 2008 J. Phys.: Conf. Ser. 100 082001
[8] Sittinger V, Szyszka B, Bandorf R, Vergöhl M, Pflug A 2008 Proc. of 51st SVC Annual TechCon 35
[9] Barankova H, Bardos L and Berg S 1995 Vacuum 46 1433
[10] Barankova H and Bardos L 2003 Surf. Coatings Technol. 163-164 649
[11] Birkholz M, Albers U and Jung T 2004 Surf. Coat. Technol. 179 279
[12] Jung T and Westphal A 1993 Surf. Coat. Technol. 59 171
[13] Bandorf R, Loch D, Bräuer G 2011 Proc. of 54th SVC Annual TechCon PAGE
[14] Yukimura K, Mieda R, Tamagaki H and Okimoto T 2008 Surf. Coatings Technol. 202 5246
[15] Azuma K, Mieda R, Yukimura K, Tamagaki H and Okimoto T 2011 Surf. Coatings Technol. 206 938
[16] Yukimura K and Ehiasarian A P 2009 Nucl. Instr. Meth. Phys. Res. B 267 1701
[17] Loch D and Ehiasarian A P 2011 J. Phys.:Conf. Ser. to be published
[18] Anders A, Andersson J and Ehiasarian A P 2007 J. Appl. Phys. 102 113303
[19] Andersson J and Anders A 2009 Phys. Rev. Lett. 102 045003
[20] Andersson J and Anders A 2008 Appl. Phys. Lett. 92 221503
[21] Anders A 2008 Appl. Phys. Lett. 92 201501
[22] Ehiasarian A P, New R, Münz W D, Hultmann L, Helmersson U and Kouznetsov V 2002 Vacuum 65 147
[23] Sarakinos K, Alami J, Klever C and Wuttig M 2007 Rev. Adv. Mater. Sci. 15 44
[24] Bandorf R, Vergöhl M, Schiffmann K and Wallendorf T 2006 Proc. of 49th SVC Annual TechCon 21
[25] Petrov I, Orlinov V, Ivanov I, Koutrev J and Jelev J 1989 Thin Solid Films 168 239
[26] Ullmann J, Wolf G K and Möller W 1993 Nucl. Instrum. Methods Phys. Res. B 80/81 1507
[27] Greczynski G, Jensen J and Hultman L 2011 Thin Solid Films 519 6354
[28] Wallendorf T, Bandorf R and Gerdes H 2011 J. Phys.:Conf. Ser. to be published

IOP Conf. Series: Materials Science and Engineering **39** (2012) 012006 doi:10.1088/1757-899X/39/1/012006

A novel sputtering technique: Inductively Coupled Impulse Sputtering (ICIS)

D A L Loch, A P Ehiasarian

HIPIMS Technology Centre, Sheffield Hallam University, Sheffield, UK

Abstract. Sputtering magnetic materials with magnetron based systems has the disadvantage of field quenching and variation of alloy composition with target erosion. The advantage of eliminating magnetic fields in the chamber is that this enables sputtered particles to move along the electric field more uniformly. Inductively coupled impulse sputtering (ICIS) is a form of high power impulse magnetron sputtering (HIPIMS) without a magnetic field where a high density plasma is produced by a high power radio frequency (RF) coil in order to sputter the target and ionise the metal vapour. In this emerging technology, the effects of power and pressure on the ionisation and deposition process are not known. The setup comprises of a 13.56 MHz pulsed RF coil pulsed with a duty cycle of 25 % . A pulsed DC voltage of 1900 V was applied to the cathode to attract Argon ions and initiate sputtering. Optical emission spectra (OES) for Cu and Ti neutrals and ions at constant pressure show a linear intensity increase for peak RF powers of 500 W - 3400 W and a steep drop of intensity for a power of 4500 W. Argon neutrals show a linear increase for powers of 500 W - 2300 W and a saturation of intensity between 2300 W - 4500 W. The influence of pressure on the process was studied at a constant peak RF power of 2300 W. With increasing pressure the ionisation degree increased. The microstructure of the coatings shows globular growth at 2.95×10^{-2} mbar and large-grain columnar growth at 1.2×10^{-1} mbar. Bottom coverage of unbiased vias with a width of 0.360 μm and aspect ratio of 2.5:1 increased from 15 % to 20 % for this pressure range. The current work has shown that the concept of combining a RF powered coil with a magnet-free high voltage pulsed DC powered cathode is feasible and produces very stable plasma. The experiments have shown a significant influence of power and pressure on the plasma and coating microstructure.

Keywords: ICIS, Ionised PVD, Magnet-free sputtering, deposition on high aspect ratio vias

1. Introduction

During sputtering charged particles i.e. ions, follow the electric field lines that are created in the chamber by the electric potential difference between the plasma bulk and the substrate surface. [1] This means that sputtering with a highly ionised metal plasma makes it possible to deposit coatings with good bottom coverage within high aspect ratio structures [2 - 4].

Conventionally, to achieve the high electron density necessary to ionise the working gas, a magnetic field needs to be applied on the target surface. To ionise the metal vapour additionally high power is necessary. Especially for sputtering magnetic materials the field is shunted and limits the usable target thickness. Furthermore, this additional magnetic field has the disadvantage of disturbing the uniform flow along the electric field lines. While this disturbance is advantageous for the initial

-----

Corresponding author.
E-mail address: daniel.a.loch@student.shu.ac.uk (D Loch)

ionisation of the gas and metal vapour as the collision probability is greatly enhanced, during the movement to the substrate the electrons are forced on cycloidical paths along the magnetic field lines and are confined near the target thus trapping gas and metal ions. [1]

Yukimura and Ehiasarian [5] have designed a magnet free experimental apparatus that consists of 2 electrodes immersed in an 200 kHz RF plasma that is produced by an inductor.

To extend the work described, we examine the ionisation degree in a plasma generated with a RF frequency of 13.56 MHz and a coil positioned within a vacuum chamber. In the current work an radio frequency (RF) powered Copper (Cu) coil is assembled in front of the target (Cu or Ti) to sputter and ionise the metal vapour. The current flowing through the coil induces a magnetic field that confines electrons to the inside of the coil thus enhancing the collision probability[6]. To attract the argon (Ar) ions to the target for sputtering, a high voltage pulsed DC is applied to the target. In the current work the effect of RF power on ionisation is studied and modelled by optical emission spectroscopy (OES) and the bottom coverage of Cu in vias is studied.

2. Experimental Details

The experimental ICIS system (schematic in Fig.1) consists of a UHV chamber (Kurt J. Lesker), a Hüttinger PFG 5000 RF power supply (13.56 MHz), a Advanced Converters HIPIMS power supply, a 2-turn 80 mm diameter solid rod copper coil and a magnet free 75 mm diameter cathode.

The plasma discharge is created within the RF powered coil. When the plasma has ignited, pulsed DC power is applied to the cathode. RF and DC power pulses are synchronised by a pulse generator.

In the current work a pressure - RF-power matrix was used to examine the influences of working pressure and RF- power on the discharge while the pulsed DC parameters were kept constant. During the experiments the working pressure was varied from 2.14×10^{-1} - 2.96×10^{-2} mbar and the RF power was varied between 500 W - 4500 W the DC voltage remained constant at 1900 V. The repetition frequency was 500 Hz with a pulse width of 500 µs, which relates to a duty cycle of 25%. The substrate was silicon oxide (SiO_2), an insulator, with vias, the bias voltage was floating. Temperature on the substrate at the beginning of the process was between 20 - 28 °C and during deposition the temperature rose by aprox. 5 °C within one hour.

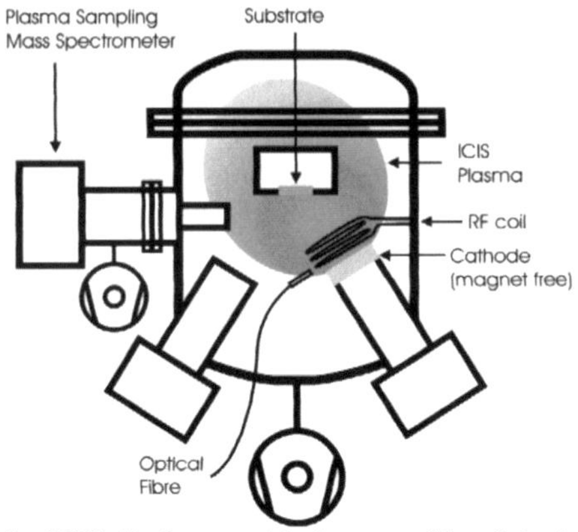

Figure 1. Experimental set up of the ICIS discharge with the assembly of the inductive coil, magnet free cathode and optical emission spectroscopy.

2.1. Plasma and coating characterisation techniques

Analysis was carried out by Optical Emission Spectroscopy (OES) (Jobin Ivon Triax 320) with quartz optical fibre and collimator *in vacuo*. To compare the ionisation performance of ICIS with the well understood conventional rf coil enhanced magnetron sputtering technique, the results of the OES measurements were fitted into a model based on magnetron sputtering based on electron ionisation collisions.[7]

Scanning Electron Microscopy (SEM) (FEI NovaSEM 200) was used to determine the bottom coverage, i.e. the ratio of the coating thickness on the bottom and top surfaces of vias.

2.2. Modelling

The used model correlates the optical emission of the plasma to the power on the cathode [7], in case of this work, it is the power applied to the induction coil. The model explains the connection between the intensity *(I(λij))* and power (P) for highly ionised magnetron plasma processes. It shows the increase in metal ionisation as a function of power. The intensity is expressed by a power law with exponent β. β is derived from the slope in a log — log graph.
The following subchapter will describe the basis of the model
The intensity of emission is expressed by the density of the components,

$$I(\lambda ij) = c(\lambda ij)\frac{Aij}{\sum_j Aij} \cdot n_0 \cdot n_e \cdot C^i \quad (1),$$

where $I(\lambda ij)$ is the intensity, $c(\lambda ij)$ is the spectral response of the spectrometer, Aij is the radiative frequency, n_0 is the gas neutral density, n_e is the electron density and C^i is the production rate by electron collision. The spectral response of the spectrometer $(c(\lambda ij))$, the radiative frequency (Aij) and the production rate by electron collisions can be assumed as constant (Kij) with respect to electron density.
For the excitation of Argon $(e + Ar \rightarrow Ar^*)$ Eq.1 can be written as follows:
$I_{Ar} = K_{Ar} \cdot n_{Ar} \cdot n_e \cdot C^{Ar}$(2) As most parts of the equation are constant $I_{Ar} \propto n_e$. Further it is assumed that C^i is constant with increasing power so that $n_e \propto P^\beta$, where P is the power and β is the slope in a log-graph. This concludes that $I_{Ar} = P^\beta$ (3).
We determine β by taking the logarithm of both sides of the equation are logarithmised to get and obtain: $\log(I_{Ar}) = \beta \log P$ (4).
Excitation of Ti^0 $(e + Me \rightarrow Me^* + e)$ is: $I_{Me} = K_{Me} \cdot n_{Me} \cdot n_e \cdot C^{Me}$ (5).
Because plasma is considered quasineutral, $n_{Ar^+} \approx n_e$ and from the definition of sputtering yield it is known that $n_{Me} = \varepsilon \gamma_e n_{Ar^+}$ (6). Where ε is a constant and γ_e is the sputtering coefficient, it can be concluded that $I_{Me} \propto {n_e}^2$. From this it follows that $I_{Me} = (P^\beta)^2$ (7) and $\log(I_{Ion}) = 2\beta \log P$ (8).
For the ionisation of Ti $(e + Me \rightarrow Me^+ + 2e)$, the intensity is $I_{Ion} = K_{Ion} \cdot n_{Ion} \cdot n_e \cdot C^{Ion}$(9). Where $n_{Ion} \approx n_{Me} \cdot n_e \approx {n_e}^2$. Thus $I_{Ion} \propto {n_e}^3$ and $I_{Ion} = (P^\beta)^3$ (10) which results in $\log(I_{Ion}) = 3\beta \log P$ (11).

3. Results and modelling

3.1. Current and voltage

The current and voltage behaviour on the cathode during a pulse can be seen in Fig. 2. The influence of the DC voltage on the current can be clearly seen, with the rise in current during the pulse. Hereby the target is biased to 1900V. Note that the voltage does not drop to zero as the current is too low to discharge the generator over the plasma.

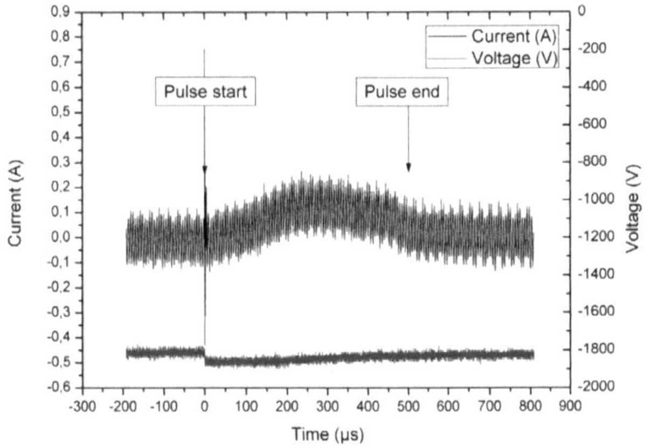

Figure 2 Current and voltage over time on the Cu-cathode during a pulse at 1.2x10-1 mbar. The target is biased to 1900V. The current induced by the impinging ions is upto 0.3A.

3.2. Optical Emission

The OES measurements in fig. 3a and 3b show the intensities of the individual species in the ICIS plasma. In both Ti and Cu cases, strong emission from single-charged metal ions is present. Ar and Cu neutrals are detected as well. Due to limited availability of spectral lines for metal ions and neutrals in the optical spectrum, lines that are visible at all powers were chosen. Emission lines were transitions to ground state with similar upper excitation levels that were generally lower than the Ar neutral excitation level of 15 eV. Spectra lines data was taken from NIST.[8]

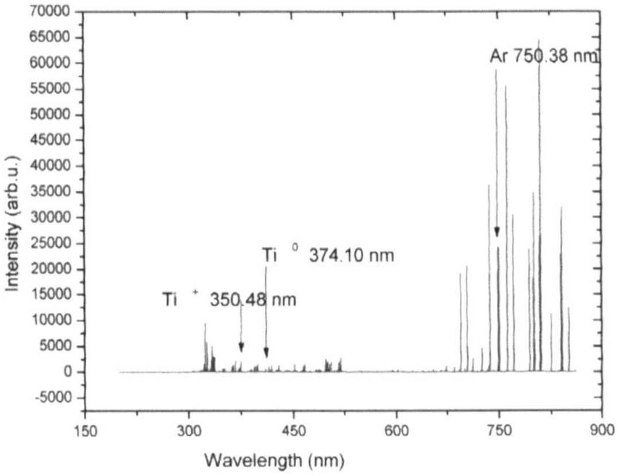

Figure 3a OES measurement of Ti ICIS plasmas at 1.2x10-1 mbar, 2300 W RF power. Lines used for analysis have been marked.

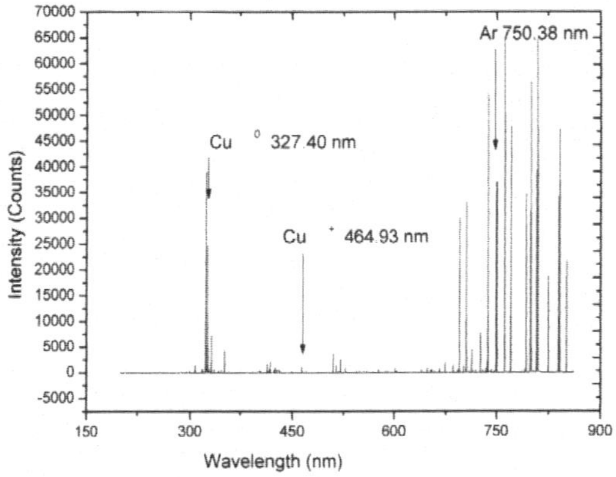

Figure 3b OES measurement of Cu and Ti ICIS plasmas at 1.2x10-1 mbar, 2300 W RF power. Lines used for analysis have been marked.

3.3. Modelling the influence of power on ionisation

Calculations from the model predict a linear rise of intensity with increasing power in a log-log graph. In the case of ICIS of Ti plasma, measurements show that the slope of Ar is 0.61±0.08. As the model does not differentiate between sputter elements this means that according to the model the slope of the Ti^0 should be 1.22±0.08 and the slope of Ti^+ should be 1.83±0.08. As the slopes of the metal species are based on the Ar slope, they all share same error margin.

Figures 4a-c clearly show that all measured values fit very well into the error margin of the slope of the model. The Ti neutrals and ions rise with increasing power, with a sharp drop for very high RF power. Ar intensity rises continuously with power.

The rate of rise of metal ions is faster than that of metal neutrals. Thus the ionisation degree increases with increasing power.

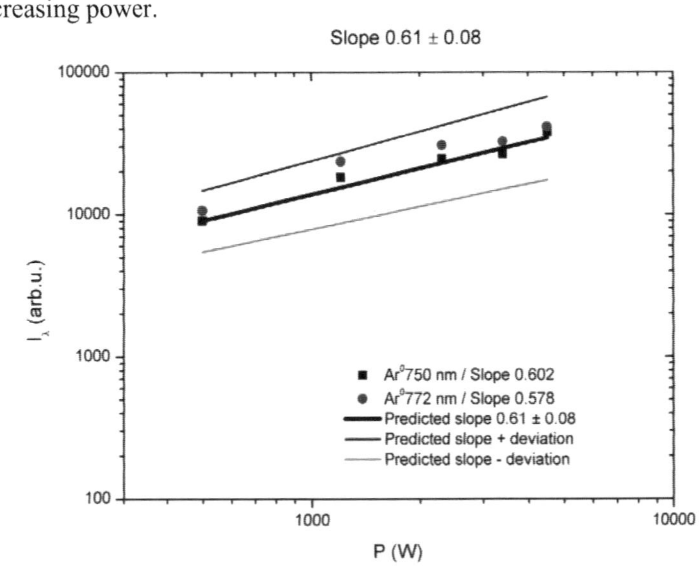

Figure 4a Measured and predicted results for the increase of ionisation with increasing power.

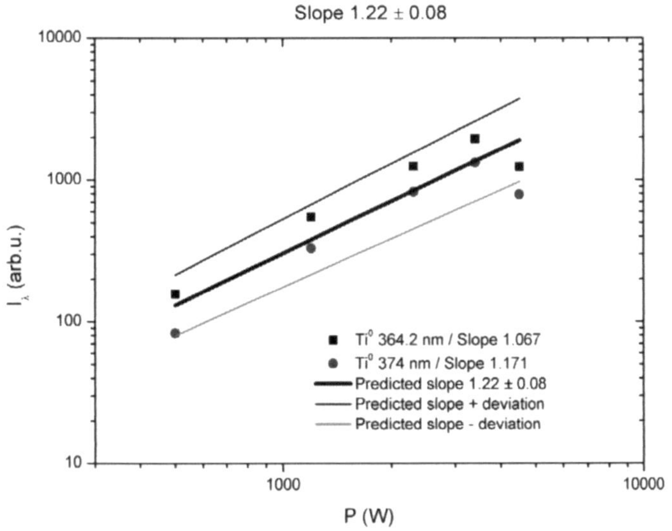

Figure 4b Measured and predicted results for the increase of ionisation with increasing power

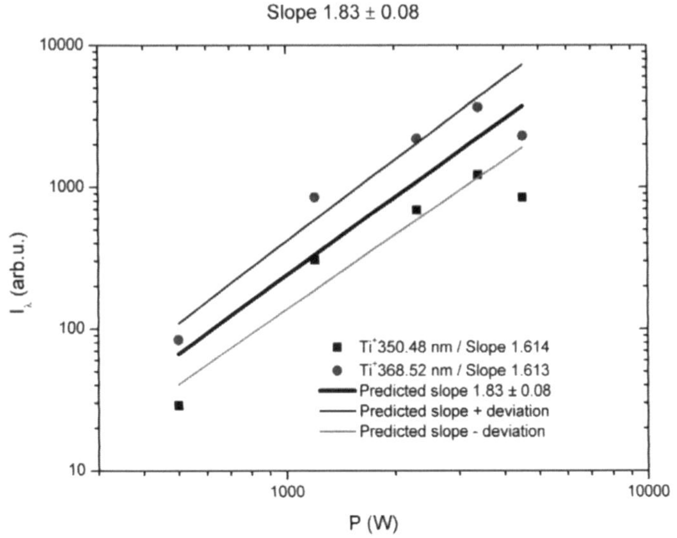

Figure 4c Measured and predicted results for the increase of ionisation with increasing power

Fig. 5 shows the increase in the ionisation degree as an ion to neutral ratio. For powers above 1000 W RF the rise is linear with increasing power. Below 1000 W the power is too low to generate Ti ions.

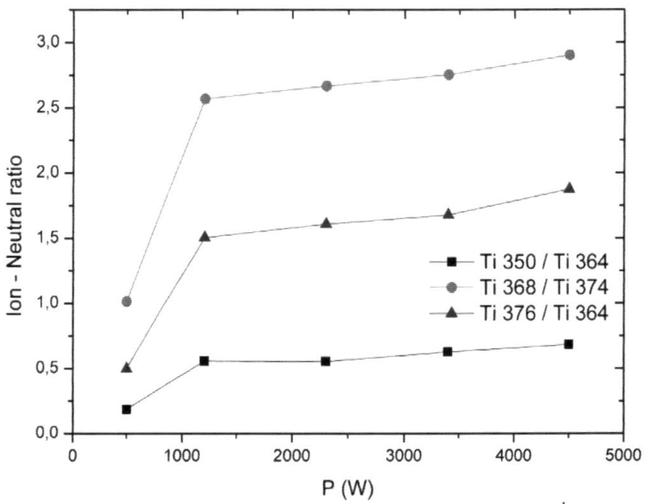

Figure 5 Ion to Neutral ratio for Ti ICIS plasma at $1.2x10^{-1}$ mbar

Fig.6 shows similar measurements for an ICIS of Cu plasma. When comparing the optical emission from ICIS of Ti and Cu plasmas it can be seen that upto a power of 2300 W both show similar behaviour for the metal species with a linear increase of ionisation as a function of RF power. For higher powers there is a saturation for the Cu metal species.

When sputtering Ti the intensity of Ar rises continuously with increasing power, while when sputtering Cu the intensity of Ar saturates for higher powers.

Table 1 lists the slopes as predicted by the model for Ti and the slopes of the measured Cu plasma species. Cu deviates from the model as the metal slopes are not clearly related to those of the Ar. Additionally, the slope for Ar is significantly greater when Cu is used. The reasons behind this need to be examined further.

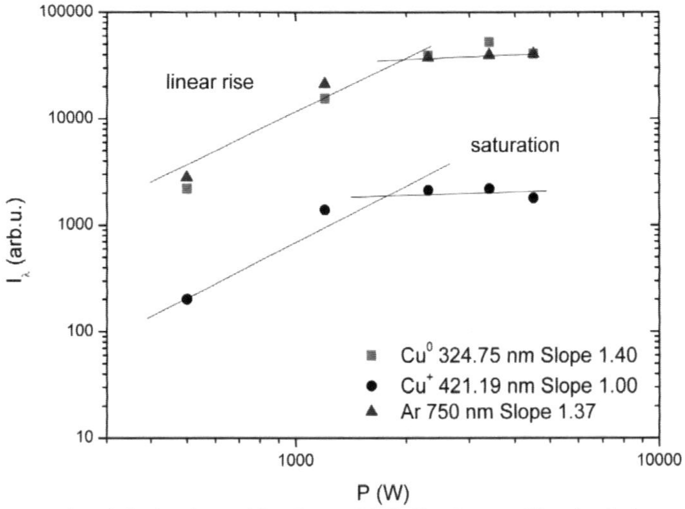

Figure 6 Ar and Cu neutral and Cu ion intensities for an ICIS Cu plasma. The rise in intensity is linear with power. Pressue $1.2x10^{-1}$ mbar.

Table 1 Comparison of the slopes of Ar and Cu plasmas. The rise in intensity is linear with power for Ti. Also the Model will need to be adapted to take into account the lower ioniastion potential of metal ions compared to argon.

Ar^0 slope:	0.61	Ar^0 slope:	1.37
Ti^0 slope:	1.22	Cu^0 slope:	1.41
Ti^+ slope:	1.83	Cu^+ slope:	1.76

To get better results for high density plasmas, the model will need to be adapted to take into account the lower ionisation potential of metal ions compared to argon. This can be seen in the results for Cu, where the measured results do not fit into the model prediction. Cu has a higher sputter rate compared to Ti and as such, probably the metal ion density is considerably higher.
Also due to the possibillity of successive excitations the model results overestimate the presence of metal ions.

3.4. SEM

The cross sectional (Fig. 6) SEM image of a high aspect ratio via coated with ICIS of Cu shows good overall coverage. The deposition process was carried out at the higher end of power and, respectively, metal ionisation degree. For this high pressure ($1.2*10^{-1}$ mbar) process the bottom coverage is 21.6 %. The deposition rate was ca. 80 nm/h at 2300 W.

Table 2 compares the bottom coverage between high and low pressure depositions. A general trend was that bottom coverage improved at higher pressures.
The microstructure was influenced strongly by the pressure as well. Dense columnar structure was observed at the higher pressure ($1.2*10^{-1}$ mbar) and dense globular growth at the lower pressure of $2.9*10^{-2}$ mbar. The deposition rate is 99-119 nmh^{-1} for RF-power of 2300 W, average target power of 67 W and a pressure of 1.2×10^{-1} mbar.

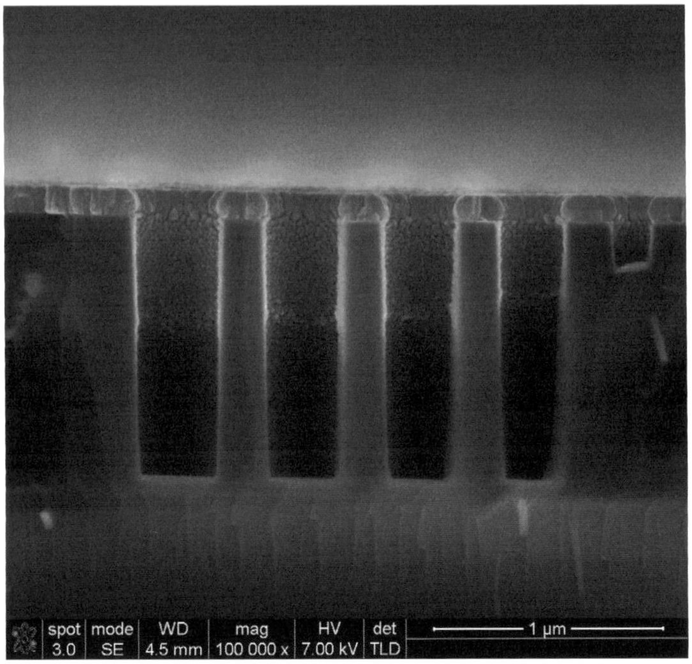

Figure 7 SEM cross section of ICIS Cu coated high aspect ratio vias at 2300 W, 1.2*10-1 mbar.

Table 2 Bottom coverage of vias obtained at high and low pressure depositions.

	Pressure:	$1.2 * 10^{-1}$ mbar	
Height	Width	Top thickness	Bottom thickness
1,40 μm	315 nm	120 nm	18nm
Ratio:	4:1	Bottom coverage:	21,6 %
	Pressure:	$2.96 * 10^{-2}$ mbar	
495 nm	143 nm	91.8 nm	15.3 nm
Ratio:	3.5:1	Bottom coverage:	14,0 %
240 nm	143 nm	91.8nm	15.3 nm
Ratio:	1.7:1	Bottom coverage:	14,0 %
163 nm	143 nm	91.8 nm	20.4 nm
Ratio:	1.2:1	Bottom coverage:	18.70%

4. Conclusion

Successful experiments were carried out by sputtering Cu and Ti that present ICIS as a feasible sputtering technique. Working without magnetic field eliminates the racetrack that is created with magnetron sputtering, thus enhancing target utilisation and allowing ions to follow the electric field lines without disturbance.

The measured results of the rate of increase for gas neutrals, metal neutrals and metal ions fit very well to the predicted values of the model for magnetron sputtering. Calculations from the model confirm the increase of the ionisation degree as a function of power for Ti. Further examination has to be done into the sharp drop in intensity for metal neutrals and ions for high RF

Power and adjustments to the model have to be made to take this and the lower ionisation potential, as well as successive exitation into account.These results show that the ionisation mechanisms for ICIS are the same, by type as well as quantity, as for RF-enhanced magnetron sputtering, but without the influence of the magnetic field of the magnetron.

Copper deposition on high aspect ratio vias has shown very promising results. The bottom coverage in vias with aspect ratio of 4:1 is 21.6 % with dense columnar growth and without bias.

References

[1] Chapman B 1980 Glow Discharge Processes: Sputtering and Plasma Etching. *Wiley-Interscience; 1 edition.*

[2] Grapperhaus M, Krivokapic Z, Kushner M J 1997 J. Appl. Phys. **83**(1), 35 - 43

[3] Hopwood J, Qian F 1995 *J. Appl. Phys.* **78** (2), 758 – 765.

[4] Rossnagel S M, Hopwood J 1993 *J. Vac. Sci. Technol.* **B 12(1)**, 449 – 453.

[5] Yukimura K, Ehiasarian A P 2009 *Nuclear instruments and methods in physics research section B: Beam interactions with materials and atoms,* **267** (8-9), 1701-1704.

[6] Hopwood J 1992 *Plasma Sources Sci. Technol.* **1**, 109 - 116.

[7] Dony M F, Ricard A, Dauchot J P, Hecq M, Wautelet M 1995 *Surf. Coat. Technol.,* **74-75**, 479-484.

[8] NIST atomic spectra database: http://www.nist.gov/pml/data/asd.cfm

IOP Conf. Series: Materials Science and Engineering **39** (2012) 012007 doi:10.1088/1757-899X/39/1/012007

Pulse Magnetron Sputtering with high power density – an attempt at a critical evaluation

Peter Frach, Christian Gottfried, Fred Fietzke, Heidrun Klostermann, Hagen Bartzsch, Daniel Glöß

Fraunhofer-Institut für Elektronenstrahl- und Plasmatechnik FEP, Dresden, Germany

Abstract. In this paper specific advantages and disadvantages of different pulse magnetron sputtering processes (unipolar, bipolar and HIPIMS) as well as current and potential fields of application will be discussed. On the examples of ZrN, Ti and TiO_2 the typical effects and their influence on film properties occurring during the transition from classical medium frequency pulse magnetron sputtering to high energy pulse sputtering will be described. The discharge current density was varied between 0.2 and 3.5 A/cm^2. Aspects of energy feed-in and reactive process control in the transition mode will be considered. Furthermore the influence of rising ionisation on the occurrence of crystalline phases and on mechanical, optical and photocatalytic properties of the layers will be presented. The paper concludes with a placement of the processes related to other PVD-processes that is based on further own experimental results and evaluation of dependencies as well as considering published results of other groups regarding pulse magnetron sputter processes of high power density for the deposition of hard coatings and TCO.

1. Introduction

Since the idea of high power pulsed magnetron sputtering (HPPMS or HIPIMS) was born more than 10 years ago [1], there has been ever increasing interest in scientific circles to use this new technology, carry out extensive R&D, and clarify the mechanisms involved [2]. In some areas, this technology has attained special technical significance, for example for "Highly Ionized Metal Sputtering" in microelectronics which is used for coating deep trenches of high aspect ratio with metals [3]. In contrast, no applications for large surfaces have been developed. In industry a certain reticence to use the technology, and to transfer the acquired knowledge to real processes, has prevailed.

HIPIMS, like conventional PMS, is concerned with feeding energy pulses into a magnetron discharge – albeit with much higher power densities and hence drastically reduced values for the pulse-pause ratio and pulse frequency. Typically conventional PMS has power densities during the pulse-on-time of between 5 ... 50 W/cm^2 for pulse-pause ratios of 1:1 and pulse frequencies of between 20 ... 200 kHz. The relevant values for HIPIMS are 0.5 ... 5 kW/cm^2, 1:1000, and 50 ... 500 Hz (Fig. 1).

Corresponding author.
E-mail address: peter.frach@fep.fraunhofer.de (P Frach); phone: +49 (0) 351-2586-370; fax +49 (0) 351-2586-55-370

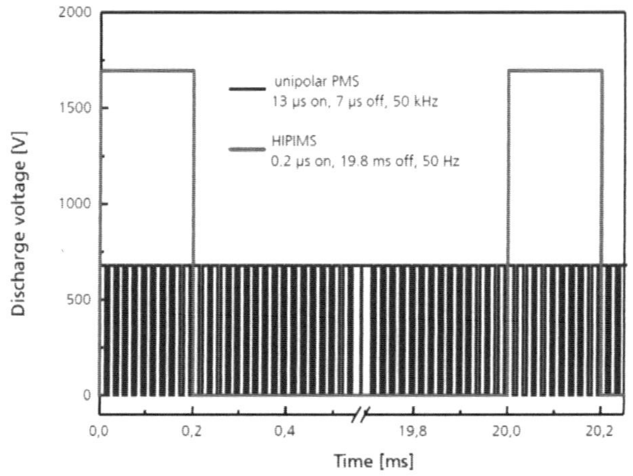

Figure 1. Schematic time profile of the discharge voltage for unipolar PMS and for HIPIMS.

The high power density in the pulse leads to a high degree of ionization of the layer-forming particles, allowing the deposition of layers which are especially dense, smooth, and homogenous. Besides this advantage, there are a number of disadvantages and in particular the drastically reduced deposition rate and the much higher tendency for arc discharges.

In order to build up a knowledge base at the Fraunhofer FEP for evaluating this technology, the HIPIMS generator TruePlasma HighPulse 4008 (20 kW DC, 2 kV / 4 kA pulse at max. 500 Hz) made by Trumpf Hüttinger Elektronik, was compared with established MF pulse generators under a wide range of conditions (coating plant, magnetron type, target material, pressure, gas composition). The discharges themselves and also the deposited layers were evaluated.

2. Results and dicussion

2.1. ZrN layers

The first tests were carried out in the batch coating plant UNIVERSA with a 120 x 500 mm^2 rectangular magnetron on the Zr-N system. The generators which were used were the HIPIMS generator and a voltage-sourced rectifier for bipolar and unipolar pulse mode made by MagPuls (60 kW DC, 1 kV / 1kA pulse at max. 50 kHz). The maximum attained power densities in the pulse were about 0.5 kW/cm^2 for the HIPIMS generator, which was a factor of 2 higher than for the voltage rectifier. This was due to the higher operating voltage. Limiting factors for the power density of both generators were the magnetron magnetic field and arcing for long pulse lengths.

In order to evaluate the properties of the layers, a bipolar pulsed standard process (20 kHz, 100 W/cm^2 pulse) was compared with a HIPIMS process (500 Hz, 0.5 kW/cm^2 pulse) for depositing ZrN anti-abrasion layers. As expected, a slight increase in the hardness and an improvement of the morphology towards denser and smoother layers were observed (Fig. 2a, b). The deposition rate per unit power on rotating substrates fell insignificantly for the power density achieved in HIPIMS mode. However, the level of contamination in the layers increased somewhat due to the shorter duty cycle.

2.2. Ti and TiO_2 layers

Further experiments with the "True Plasma" HIPIMS generator were conducted using the inner target of the double ring magnetron sputter source DRM 400 on the Cluster 300 plant [4,5]. Amongst the materials which were studied, comparative tests were carried out with a Ti target with and without reactive gas. The DRM 400 has moveable magnet systems so that the dependence of the energy input in HIPIMS mode on the magnetic field strength could be studied in detail. It was found that the power density on the target was very dependent on the magnetic field strength for sputtering in pure argon (Fig. 3).

Figure 2a. SEM micrographs of ZrN layers deposited via bipolar PMS.

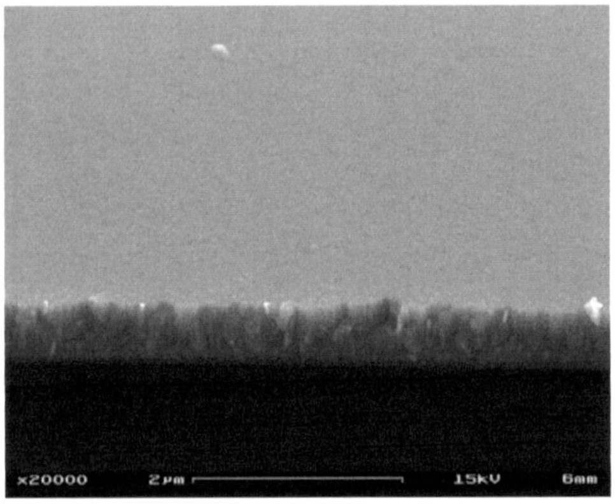

Figure 2b. SEM micrographs of ZrN layers deposited via HIPIMS.

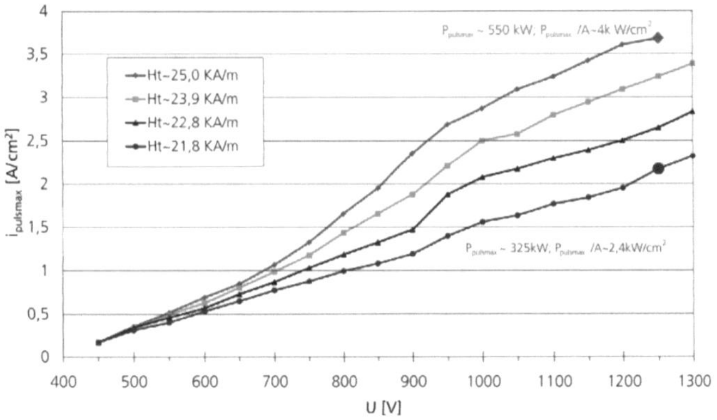

Figure 3. Maximum discharge power density $i_{pulsmax}$ in the pulse for a Ti target (magnetron DRM 400) as a function of the discharge voltage U at different magnetic field strengths (p = 1 Pa Ar, f = 12 Hz, t_{on} = 200 μs).

The maximum power density in the pulse increases at the target from 2.4 kW/cm^2 to 4 kW/cm^2 (67% increase) when the magnetic field strength is only increased by 15%. To achieve high power densities, strong magnetic fields are therefore favorable. The Ti coating rates in HIPIMS mode are only 10 to 14 percent of the coating rates in unipolar pulse mode at the same average power (Tab. 1). Furthermore, it was found that the thermal load on the substrate per unit layer thickness (namely the temperature increase after depositing a layer of defined thickness) was comparable with that of the PMS process.

The following are results from experiments in reactive HIPIMS mode with a Ti target in oxygen: In HIPIMS mode, the reactive process also showed hysteresis. This contradicts other published findings. The reactive HIPIMS process can be controlled using optical plasma emission spectrometry.

Higher process stability was achieved using the plasma control unit S-PCU of the Fraunhofer FEP by using the ratio of the intensities of 2 spectral lines (Ti 500 nm; O_2 777 nm). By controlling the process in this way, the discharge can be stabilized within any desired working points and stoichiometric TiO_2 layers can be deposited in both unipolar pulse mode and in HIPIMS mode.

The maximum achievable coating rates for stoichiometric TiO_2 are only 5 to 20 percent of the comparable rates in unipolar mode. The TiO_2 layers deposited in the HIPIMS mode are much finer grained than those deposited by unipolar PMS. They grow from the substrate, and are first of all amorphous and then in the crystalline rutile form. Similar behavior was observed for TiO_2 layers deposited in bipolar pulse mode or in pulse packet mode. The TiO_2 layers deposited in the HIPIMS mode have, however, a higher hardness of up to 20 GPa compared to a maximum of 16 GPa for those deposited in pulse packet mode. They have a somewhat higher refractive index (n) than the layers deposited in unipolar pulse mode [6].

Table 1. Summary of the layer properties for different pulse modes and different reactive working points for TiO2 deposition.

mode	working point	layer thickness [nm]	rate [nm/min]	hardness [GPa]	n 550 nm	k @550 nm
unipolar	stoichiometric	560	110.0	~7.0	2.480	$< 10^{-5}$
HIPIMS	hypostoichiometric	848	28.0	20.0	2.552	4.70×10^{-3}
HIPIMS	full reactive	180	6.0	16.5	2.514	4.20×10^{-3}
HIPIMS	stoichiometric	675	18.4	19.1	2.565	2.95×10^{-3}

3. Summary

The studies on ZrN and TiO2 have shown that HIPIMS is a promising technology for achieving layer properties that cannot be achieved using other sputtering techniques. These properties include a higher hardness and fine crystallinity. The disadvantages of the technique are in particular the much reduced coating rate at higher power density and the process stability. Given the current state of HIPIMS generators, laboratory scale process development should ensue. For the further development of the technology, suitable magnetron sputter sources with a high magnetic field and power supplies with sufficient performance to allow upscaling of the process are required. This will then enable HIPIMS processes to be developed for specific applications.

References

[1] Vladimir Kouznetsov, Karol Macak, Jochen M. Schneider, Ulf Helmersson, Ivan Petrov: A novel pulsed magnetron sputter technique utilizing very high target power densities, Surface and Coatings Technology 122 (1999) 290–293

[2] K. Sarakinos, J. Alami, S. Konstantinidis: High power pulsed magnetron sputtering: A review on scientific and engineering state of the art, Surface and Coatings Technology 204 (2010) 1661–1684

[3] Ulf Helmersson, Martina Lattemann, Johan Bohlmark, Arutiun P. Ehiasarian,Jon Tomas Gudmundsson: Ionized physical vapor deposition (IPVD): A review of technology and applications, Thin Solid Films 513 (2006) 1–24

[4] H. Bartzsch, P. Frach *, K. Goedicke, Chr. Gottfried: Different pulse techniques for stationary reactive sputtering with double ring magnetron, Surface and Coatings Technology 120–121 (1999) 723–727

IOP Conf. Series: Materials Science and Engineering **39** (2012) 012007 doi:10.1088/1757-899X/39/1/012007

[5] Hagen Bartzsch, Peter Frach, Klaus Goedicke: Abscheidung optisch, elektrisch und akustisch wirksamer Schichten mit dem Doppel-Ring-Magnetron, Vakuum in Forschung und Praxis 15 (2003) Nr. 3 122–126
[6] M. Vergöhl, S. Bruns: Properties of TiO2 Films Deposited by Bipolar Reactive HiPiMS, ICCG 8, Braunschweig 2010 Proceedongs 301-305

IOP Conf. Series: Materials Science and Engineering **39** (2012) 012008 doi:10.1088/1757-899X/39/1/012008

Modelling of target effects in reactive HIPIMS

Tomas Kubart and Joakim Andersson

Solid State Electronics, The Angstrom laboratory, Uppsala university, Box 534, 751 21 Uppsala, Sweden

Abstract. In reactive High Power Impulse Magnetron Sputtering (HIPIMS) of oxides, target effects such as reduced surface oxidation during pulse off time, increased implantation of reactive gas due to the higher discharge voltage as compared to normal DC sputtering, and enhanced target cleaning during on time are considered to be responsible for the differences compared to reactive DC sputtering. These effects are assumed to cause changes in the target oxide coverage and hence lead to the hysteresis shifts observed in experimental studies. In this contribution, the target processes are simulated using the binary collision approximation code TRIDYN. Using an Al target sputtered in Ar+O_2 mixture as a model system, a range of pulse configurations is simulated for different oxygen partial pressures. The results indicate that the target effects alone are not sufficient to explain the observed shift of hysteresis and its frequency dependence.

1. Introduction

Formation of compound at a metal sputtering target during reactive magnetron deposition is an important effect with strong impact on the process behaviour. This effect, usually referred to as target poisoning, is demonstrated by the hysteresis effect and leads to substantial decrease in the deposition rate. The deposition rate is especially low for oxides due to the difference in sputtering yield between most metals and their oxides [1].

Wallin and Helmersson [2] showed that the hysteresis behaviour during reactive HIPIMS may differ significantly from direct current magnetron sputtering (DCMS) of the same material. In their study, the hysteresis loop was reduced or even removed during HIPIMS and this has also been confirmed by several other reports. A later study using the same setup as in ref. 2 led to the conclusion that there is an optimum frequency of pulsing for which the hysteresis loop is minimized [3]. One of the possible effects suggested to be responsible for this behaviour is the dynamics of target poisoning and cleaning due to the pulsed nature of HIPIMS [2]. The high power during a HIPIMS pulse may lead to rapid removal of the surface oxide thus increasing the effective sputtering yield. In addition, there is no implantation of reactive gas during off time which may also reduce the formation of oxide.

In this work, the sputtering and oxidation of a magnetron target is simulated using the dynamic binary collision approximation (BCA) Monte Carlo code TRIDYN. Substantial simplifications have to be made to make the model feasible. The aim is to take into considerations the most important effects which influence the target surface and analyse the target poisoning and sputtering.

2. Sputtering model

A typical HIPIMS pulse considered in this work is shown in figure 1. The discharge current reduction after the peak is related to the gas rarefaction and for longer pulses, figure 1(a), the current is substantially reduced as a result of lower gas density in the discharge region [4]. In reactive HIPIMS of oxides, short on-times are typically used in order to reduce arcing, figure 1(b). Therefore, the on

-----

Corresponding author.
E-mail address: tomas.kubart@angstrom.uu.se (T Kubart)

time of 50 μs is assumed in this study and it is assumed that Ar and O ions are the dominant sputtering species.

In the simulations, sputtering of Al in Ar+O_2 atmosphere is considered. Constant value of the discharge current density during the on time is assumed, figure 1(b). One HIPIMS cycle thus consists of two parts; first (on time) when the peak current is applied to the target and the ion beam consists of energetic Ar and O_2 ions, together with thermal O_2 from the working atmosphere. Then, during the second part (off time), only thermal O is present in the beam. Two values of the discharge voltage of 500 and 1000 V are considered in the simulations corresponding to the typical HIPIMS range of voltages. Even though the ion energy influences the implantation depth, the effect is not important in this work as shown later.

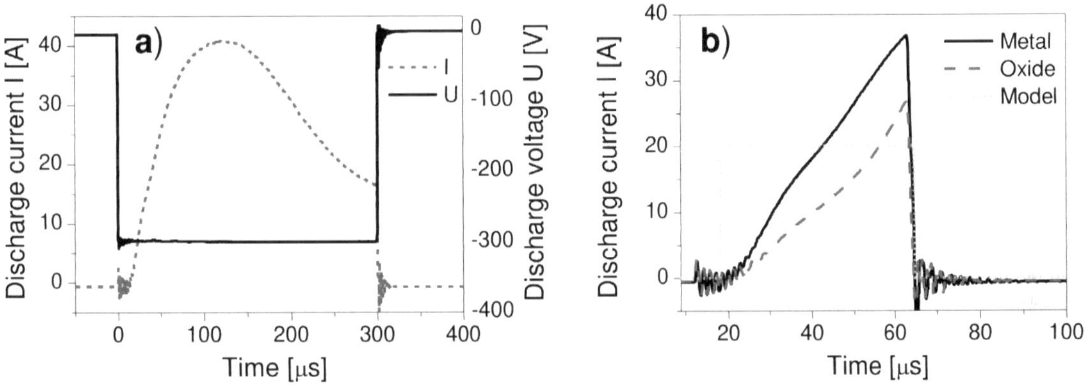

Figure 1: (a) Typical HIPIMS current and voltage waveforms during long on time, showing an initial peak in the discharge current. (b) Short pulse of 50 μs in metal and oxide mode together with the current waveform assumed in the model. For these demonstration curves, a Ti sputtering target was used. Al targets behave in a similar way.

The sputtering of a magnetron target is described using TRIDYN [5]. TRIDYN is a BCA code but unlike the popular program TRIM, TRIDYN takes into account changes in the surface composition due to the sputtering and can therefore describe the evolution of surface composition during pulses. The model takes into account O sorption on the target surface together with implantation of reactive gas. O_2 molecules are assumed to dissociate immediately upon the impact and are therefore simulated as a flux of O atoms with half of the incident energy [6]. The ratio between Ar^+ and O^+ equals the ratio of their respective partial pressures assuming comparable ionization cross-sections. Flux of neutral oxygen is calculated from the ideal gas equation for a given oxygen partial pressure. In TRIDYN, the sticking of oxygen is treated in a simplified way, assuming a unity sticking coefficient. Oxygen is therefore incorporated very efficiently into the surface and the model may overestimate oxygen chemisorption. The oxygen concentration in the target is limited to 60 at.% with the excess atoms diffusing towards the surface.

A crucial input parameter which has a strong influence on the sputtering yield is the value of the surface binding energy (SBE). In this study, a SBE dependent on the actual surface composition is used for Al with the value of 3.36eV for metallic Al and 12eV for oxidised Al based on fitting to experimental results [7, 8]. A constant value of oxygen SBE of 9eV is used, assuming the oxygen is always bonded to metal. To achieve a steady-state composition, at least 100 pulses are always simulated, using results of one pulse as the initial conditions for the following one. In the simulations, each pseudoprojectile represents differential fluence of $1.1 \cdot 10^{11}$ cm^{-2}.

It is possible to include also self-sputtering into the model by assuming a certain fixed fraction of sputtered metal to return to the target as ions. Another modification may be to divide the on time into several parts thus following better the plasma evolution. Here, we report only the simplest case assuming constant discharge current during on time and constant composition of the particle flux to the target which consists only of Ar and O.

Ar pressure of 0.5 Pa and average discharge current of 0.2 A/cm^2 are assumed. High average discharge current is used here in order to provide an upper estimate of the extent of the sputtering effects. All simulations were performed for 50 μs on time and 450, 1950, and 9950 μs off time, respectively, Table 1. For comparison, also continuous DC operation was simulated.

Table 1: Summary of pulse parameters for performed simulations.

Frequency [Hz]	100	500	2 000	DC
Peak current [A/cm^2]	40	8	2	0.2
Duty cycle [%]	0.5	2.5	10	100

3. Results and Discussion

The simulated Al sputtering yield versus the pressure of O_2 is shown in figure 2. Due to the formation of oxide at the target surface, the partial sputtering yield of Al is decreasing with increasing oxygen pressure. The sputtering yield in pure Ar is independent of the discharge mode as expected. For the lowest frequency of pulsing (long off time and high peak current) the sputtering yield in reactive mode is the highest reaching about 30% higher values than DCMS above 10 mPa. In the same region, sputtering yields for 2 kHz and 500 Hz are comparable, leading to about 20% higher value than DCMS. Both voltages lead to comparable results, at 1 000 V the target is oxidized at higher oxygen pressure as a result of higher sputtering yield at higher voltage. Note that the simulations correspond to constant current operation, and higher voltage therefore leads to higher discharge power.

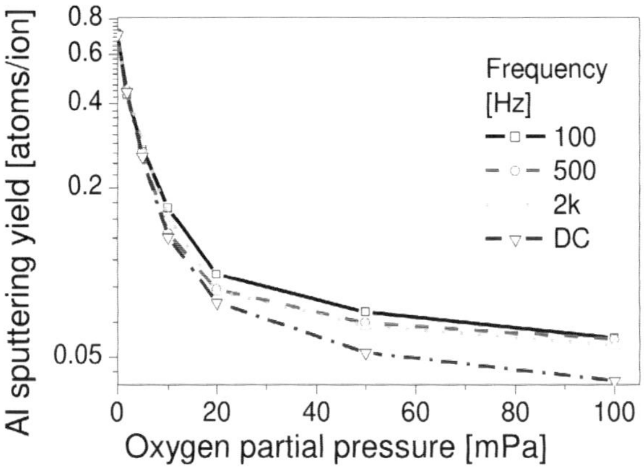

Figure 2: Partial Al sputtering yield as a function of oxygen partial pressure for different frequencies of pulsing.

In order to understand the hysteresis, target behaviour in the transition area is most interesting. Reduced oxidation would lead to less oxide on the target surface for a given O_2 partial pressure demonstrated by higher average metal sputtering yield. Therefore the evolution of surface concentration of oxygen during one cycle is studied in detail.

The results are shown in figure 3 for three different O_2 pressures showing the surface concentration in a 0.2 nm thick surface layer. In the model, the pressure of 5 mPa corresponds to an operation in the metal mode, with mainly metallic target surface, 20 mPa lies inside the transition zone and for 50 mPa the target is mostly oxidised. During the on time, the surface oxygen concentration is reduced which leads to an increase in the target erosion rate. However, even for the highest peak current used in this simulation, 40 A/cm^2, the oxygen concentration is reduced relatively little. For the given conditions, more than five monolayers of metal are removed in the metal mode and therefore a more pronounced change in the oxygen concentration would be expected. To understand this behaviour, simulated depth profiles of O are shown in figure 4 for different times within one pulse. Most pronounced change in the concentration occurs at an O_2 pressure of 20 mPa. Although the thickness of the implanted oxide layer is about 3 nm in simulations only the topmost part of about 0.3 nm changes during on time.

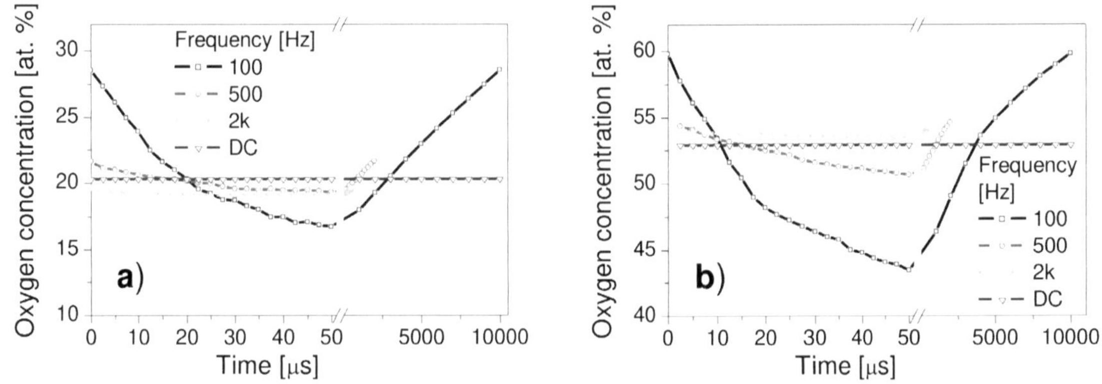

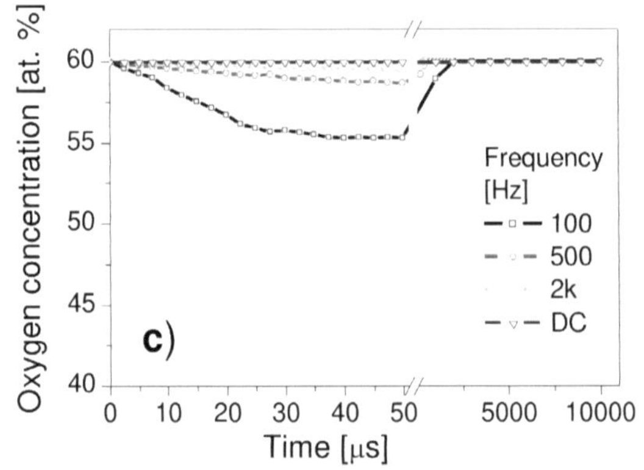

Figure 3: Simulated target surface concentration of oxygen during one cycle. The discharge voltage of 500 V was assumed and oxygen pressures of 5 (a), 20 (b) and 50 (c) mPa. The largest change occurs inside the transition region for 20 mPa. Note different time scale during on and off time.

This surprising result may be explained by the low sputtering yield of the oxide in combination with recoil implantation of O [9]. O chemisorbed at the surface is implanted deeper into the target, rather than sputtered away and the removal rate is very slow. Such behaviour was observed experimentally during XPS surface analysis employing ion etching [7].

For pressures below 20 mPa, relatively long time is required to reach the maximum surface O concentration. Even in the absence of the implantation of reactive gas during off time, O is adsorbed at the surface and implanted into the target during pulse on. Therefore, the target oxidation is not substantially limited during off time even in the model with very high sticking probability.

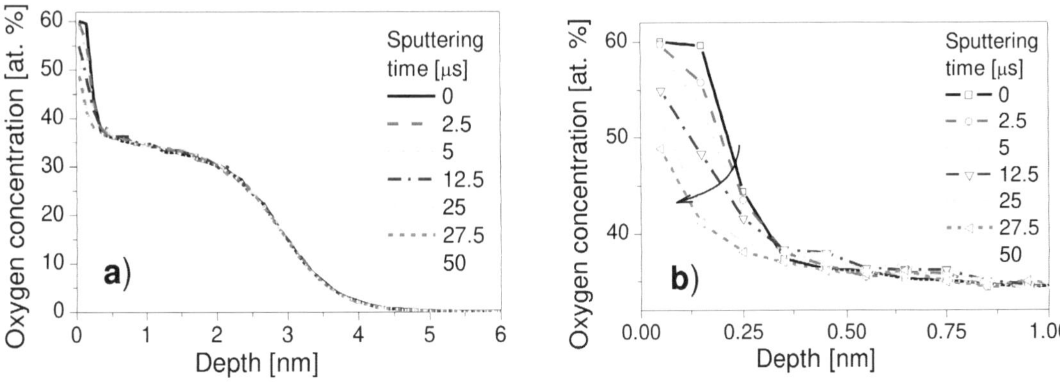

Figure 4: (a) Evolution of oxygen compositional depth profile with sputtering time during one pulse at 20 mPa and 100 Hz. (b) Detail of the topmost 1 nm showing that only a layer of 0.3 nm (about 1 monolayer) is affected during the pulse.

The simulations suggest that the O coverage decreases with increasing peak current and therefore higher deposition rate is predicted for low frequencies. Hysteresis would therefore be most reduced for longest off times and the minimum in the hysteresis observed for intermediate frequencies cannot be explained by this model. Therefore, the target effects alone are not sufficient to explain the observed shift of hysteresis and mainly its frequency dependence. In this work, very high discharge current density has been used, even higher than the common experimental values, and the effect of target cleaning will therefore be lower in reality. To our knowledge, the only effect which may explain the variations in hysteresis relates to dynamic phenomena in the gas phase, such as gas rarefaction [10]. More work is necessary to provide a conclusive answer, however.

4. Conclusions

A model has been developed to describe the sputtering of a magnetron target during pulsed operation. Simulations performed for Al target sputtered in $Ar+O_2$ atmosphere show that neither the target cleaning nor the oxygen chemisorption during HIPIMS pulses can explain the experimental observation of hysteresis in reactive HIPIMS.

5. Acknowledgements

This work was funded by Volkswagen foundation. COST action MP0804 is acknowledged for fruitful discussions.

6. References

[1] W.D. Sproul, D.J. Christie, D.C. Carter, Thin Solid Films 491/1-2 (2005) 1.
[2] E. Wallin, U. Helmersson, Thin Solid Films 516/18 (2008) 6398.
[3] M. Aiempanakit, T. Kubart, P. Larsson, K. Sarakinos, J. Jensen and Ulf Helmersson, Hysteresis and process stability in reactive high power impulse magnetron sputtering of metal oxides, Thin Solid Films (2011) doi:10.1016/j.tsf.2011.06.021
[4] A. Anders, J. Andersson, A. Ehiasarian, J App. Phys. 102/11 (2007).
[5] W. Moller, W. Eckstein, J.P. Biersack, Computer Physics Communications 51/3 (1988) 355.
[6] W. Moller, D. Guttler, J App. Phys. 102/9 (2007).
[7] T. Kubart, et al., J. Phys. D 43/20 (2010) 205204.
[8] T. Kubart, T. Nyberg, A. Pflug, M. Siemers, M. Austgen, D. Koehl, M. Wuttig, S. Berg, Surf. Coat. Technol. 204/23 (2010) 5.
[9] D. Depla, R. De Gryse, Surf. Coat. Technol. 183/2-3 (2004) 184.
[10] T. Kubart, M. Aiempanakit, J. Andersson, T. Nyberg, S. Berg, U. Helmersson, Surf. Coat. Technol. 205 (2011) S303

Reactive Sputter Deposition of Alumina Coatings

H Gerdes, R Bandorf, D Loch, G Bräuer

Fraunhofer Institute for Surface Engineering and Thin Films, Bienroder Weg 54 E, Braunschweig, Germany

Abstract. Highly ionized pulse plasma processes (HIPP processes) like High Power Impulse Magnetron Sputtering HIPIMS and Modulated Pulse Power MPP have matured in recent years. Current research focuses on the development for industrial processes. HIPP processes offer a tool for tailoring the film properties and to improve hardness, density, refractive index, and further properties beyond state of the art. Alumina coatings are used besides application in cutting tools as insulator for electric and sensor applications. This paper focuses on the process development of an industrial process for deposition of alumina with improved properties regarding the use as insulator. Concerning productivity a high rate deposition process is required for economic production. Therefore, the deposition rate must be increased or the film properties improved in a way that thinner films exceed reference films prepared with state of the art technology. MPP in comparison to a mid-frequency (MF-) process was investigated and the resulting rates are reported. The films were prepared without feedback control in the transition regime close to the threshold of the oxide mode. Regarding the insulating properties the films were characterized by their critical leakage field strength. MPP films with less than 1 μm thickness showed a critical leakage field strength up to 2.9 V/nm. The SEM cross sections of the prepared films showed a dense glassy structure for all the HIPP films.

1. Introduction

Highly ionized pulse plasma processes (HIPP processes) are investigated in the recent decade due to their potential for significant modification of the resulting film properties [1-6]. Besides well-known high power impulse magnetron sputtering (HIPIMS) significant interest is given to modulated pulse power sputtering (MPP) due to reports on film modification by ionized sputtering in combination with promising high deposition rates. First results were already published five years ago by Chistyakov et al. showing improvement in the field of protective and tribological coatings [7, 8]. Especially for reactive MPP sputtering of alumina coatings high deposition rates of approximately 12 μm per hour and a glassy structure shown by SEM investigations are published [9]. These features show high potential for use as dense insulating coating in electrical and sensor applications. Concerning industrialization of the reported process a modified setup is required, since the published films were produced using a specific setup not suited for production. Replacing the static deposition by a circular target (diameter: 17.5 cm) and a gas inlet significantly separated from the target, small industrial size rectangular targets with 322 cm^2 size and a dynamic deposition (sample rotation) were chosen for the investigations.

Corresponding author.
E-mail address: holger.gerdes@ist.fraunhofer.de (H Gerdes)

2. Experimental

The experiments were carried out in a Balzers BAS 450 sputtering plant, located in a cleanroom. The base pressure before the deposition was below $1x10^{-3}$ Pa and all deposition were carried with a constant flux of 140 sccm of Ar and a constant O_2 depending on the process parameters, which results in a working pressure of 0.5 Pa.

The plant was equipped with a PE5000 Advanced Energy mid-frequency power supply and a Z-Pulser Solo 360 power supply. Two cathodes were installed in the machine. One is used for unipolar sputtering while the second was used to close the magnetic field to realize closed field configuration for the process. The size of the cathode was 254 mm x 127 mm. All experiments were carried out without a feedback control and therefor the films were prepared in the oxide mode.

One of the key parameter for the quality of alumina thin-films as insulator is the critical leakage field strength . Therefore the films were characterized using a specific test rig for measurement of the critical leakage field strength. For the investigation an increasing voltage was applied to the deposited films until the current flow exceeded 0.5 mA, defining the voltage at 0.5 mA leakage current as the characteristic critical leakage voltage. To increase the efficiency of detecting pinholes in the layer the contact area between the tip applying the voltage and the layer was increased by using additionally a conductive liquid (saturated sodium chlorite solution).

3. Results

The investigated insulating films were prepared by unipolar MPP processes using different pulsing sequences, oxygen flow rates, and average power. Unipolar reactive MF-sputtered films served for the investigation as reference.

3.1 Mid-frequency reference process

For reference a reactive mid frequency sputtering process was chosen. Since the MPP generator did not offer the possibility of a feedback control of the target voltage, the reference MF-process was also driven without using feedback control. Another possibility to overcome the drawback of the voltage regulation for a feedback control is to control the oxygen flow by a fast piezo valve, but this option was not available for the used sputtering plant.

For the MF-process a power of 1600 W, which was in a comparable range to the later discussed MPP processes and an oxygen flow of 20 sccm resulting in a partial pressure of 2 mPa, was chosen. The resulting deposition rate was 360 nm/h. In figure 1 a SEM micrograph of a cross section of a MF-sputtered alumina reference is shown. The thin film shows a dense glassy structure. The critical leakage voltage was about 620 V for a 360 nm thick film measured with the conductive liquid.

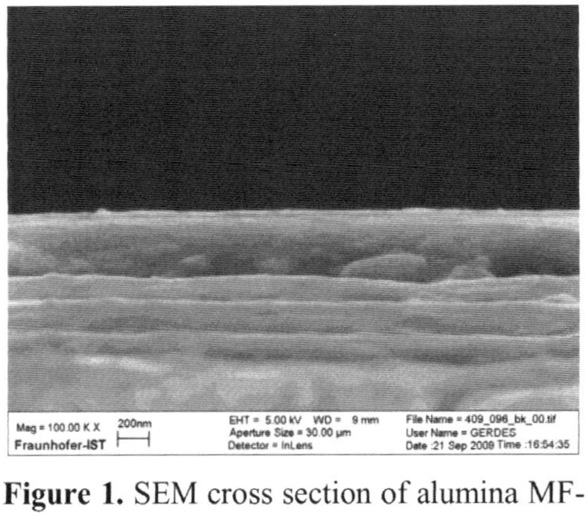

Figure 1. SEM cross section of alumina MF-reference coating; P_{ave}: 1600 W; $U_{critical\ leakage\ voltage}$ (360 nm): 620 V; 360 nm/h.

3.2 Modulated pulse power processes

The investigations for the MPP process were carried out in two steps. First a pulse sequence with only three segments was chosen. Therefore, the transition between the low and high ionisation regime was very fast and the process led to increased arcing. To avoid this instability in the transition from low to high ionisation a pulse sequence was built up, which included in total four segments. This led to slower increase of the current from the low to the high ionisation state of plasma.

3.2.1 MPP with fast transition. For the first experiment a pulse sequence with three segments was investigated resulting in a two current level state plasma. The first level was a low ionisation phase and the second a high ionisation state. The transition between these two states was very fast. The first pulse sequence is shown in table 1. The presented samples were coated using an average power of 1.7 kW and a partial oxygen pressure of 10 mPa. In comparision to the MF-process the partial pressure is higher, because of the pulsed plasma. In the off-time no significant reaction with oxygen occurs. In comparison to the MF-process in the on-time more aluminum is sputtered and therefor more oxygen is necessary to get a stoichiometric film. The alumina films were transparent and the composition was determined by EPMA (electron probe micro analysis). The composition for all films were in the range of the measurement accuracy. The extinction coefficient was not determined, although it would give a more precise feedback on the composition.

Table 1. MPP process - pulse sequence parameter with a fast transition

Segment No.	Segment on time (μs)	Segment off time (μs)	Segment duration (μs)
1	8	34	168
2	7.9	36	1273.1
3	10	15	1000

In figure 2 a SEM cross-section of the MPP film with sharp transition is shown. The thin film layer shows a dense structure with a slight columnar nearly glassy structure. This feature is also represented by the critical leakage voltage, which is about 1.4 kV for a 960 nm thick thin film layer. In comparison to the MF process the normalized critical leakage field strength (V per nm) is even lower but due to the fact that the deposition rate (960 nm/h) is higher than for the mid-frequency process, there is an increase in efficiency. For reaching the same critical leakage voltage for a thin film the deposition time is already reduced by a factor of two.

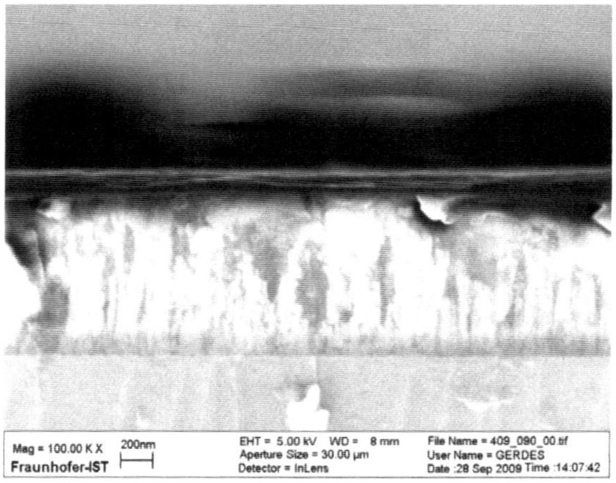

Figure 2. SEM cross section of alumina MPP coating with a fast transition from the low to the high ionisation regime;
P_{ave}: 1700; $U_{critical\ leakage\ voltage}$(960 nm): 1.4 kV; 960 nm/h

3.2.2. MPP with a slow transition. During deposition arcing was observed. From the oscilloscope it was obvious that most of the arcs occurred in the transition range between the low and the high ionisitation state. To stabilize the process a pulse sequence with a smoother transition between the low and high ionisation state was used. The parameters of the pulse sequence are shown in table 2. The resulting average power was 2.0 kW, the partial oxygen pressure 10 mPa, and the deposition rate 795 nm/h. The loss in deposition rate in comparison to the first pulse sequence is due to the choice of the working point on the hysteresis loop. This time a dense structure without any columns was observed (figure 3). In correlation with the obviously denser microstructure also the critical leakage voltage was improved by 1600 V for a 560 nm thick layer. It is assumed that the reported critical leakage field strength, higher than reported by other authors [10], is due to densification and change growth kinetics by the increase amount of ions during film formation. Already from other investigations it is known, that HIPP-process can significantly increase the film density and change the film properties compared to DC like processes [6]. Basing on these results it is possible for reaching a defined critical leakage voltage to use a thinner coating and therefore reduce the required process time significantly.

Table 2. MPP process - pulse sequence parameter with a slow transition

Segment No.	Segment on time (µs)	Segment off time (µs)	Segment duration (µs)
1	8	34	168
2	8	40	1680
3	9	25	68
4	10	15	1000

1. 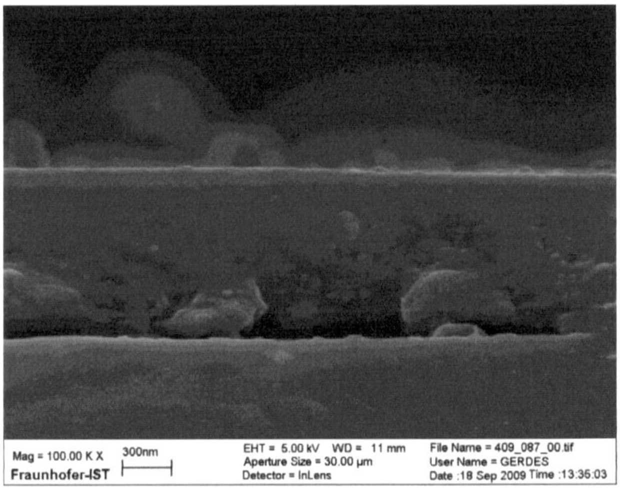

Figure 3. SEM cross section of alumina MPP coating with smooth transition between low and high ionisation regime; P_{ave}: 2000; $U_{critical\ leakage\ voltage}$(560 nm): 1.6 kV; 795 nm/h

3.3. Impact for Industrial Application

Alumina coatings used in industrial applications as insulating layer are specified by their critical leakage voltage. Therefore for a given process a defined coating thickness is required. For efficient production either a thin coating with a high critical leakage voltage, even with reduced deposition rate or a thicker coating with higher deposition rate is suited. Especially reducing the final coating thickness will further extend the maintenance intervals, saving additional time and expenses.

For the investigations carried in this paper, the resulting deposition time for layer, which withstand a voltage of 1000 V is shown in table 3. For the MF process a deposition time of 90 minutes is expected. Although the critical leakage field strength is lower for the MPP process with a fast transition between the two plasma states the resulting deposition time is reduced by a factor of 2, because of the higher deposition rate. The deposition time is about 42 minutes. The fastest process is given by the MPP with the slow transition between the two plasma states. Therefore a deposition time of 24 minutes is necessary to deposit an alumina layer which can withstand a critical leakage voltage of 1000 V. This means a reduction of deposition time by a factor of more than 3.

Table 3. Comparison of the results and calculated deposition time for a layer with a critical leakage voltage of 1000 V

	Rate (nm/h)	critical leakage field strength (V/nm)	Deposition time for a layer with a 1 kV critical leakage voltage (minutes)
MF Process	360	1.9	90
MPP Process without transition	960	1.5	42
MPP Process with transition	795	2.9	24

4. Summary

Reactive sputter deposited alumina coatings for use as insulator are of great interest, since with conventionally rf-sputtered layers have a very low deposition rate. These presented results show an opportunity for a reduction of the deposition time with unipolar setup without a feedback control. All layers were deposited dynamically. The discussed MF reference process used an average power of 1.6 kW resulting in a deposition rate of 360 nm/h and a critical leakage field strength of 1.9 V/nm. To increase the efficiency HIPP processes were investigated. MPP processes with a fast transition between the low and high ionized state of the plasma already showed an increased deposition rate of 960 nm/h at an average power of 1.7 kW. The resulting critical leakage field strength was of 1.5 V/nm. The deposition time for a coating withstanding 1000 V was already half the time needed in the MF-reference case. Even shorter deposition times can be achieved for MPP with a slow transition between the two plasma states, since the critical leakage voltage was further improved. MPP process are exhibiting promising results for alumina coatings. The depositon rate and the critical leakage field strength were higher in MPP compared to the MF process, which was used as reference.

5. Acknowledgements

The authors gratefully acknowledge the financial support of this work by the Federal Ministry of Economics and Technology (BMWi) within grant 16IN0702 and the COST Action MP0804 "Highly Ionized Pulse Plasma Processes – HIPP Processes" for networking and fruitful discussions.

6. Reference

[1] Helmersson, U.; Lattemann, M.; Bohlmark, J.; Ehiasarian, A. P. & Gudmundsson, J. T. (2006), 'Ionized physical vapor deposition (IPVD): A review of technology and applications', Thin Solid Films 513(1-2), 1-24.

[2] Alami, J.; Eklund, P.; Emmerlich, J.; Wilhelmsson, O.; Jansson, U.; Hogberg, H.; Hultman, L. & Helmersson, U. (2006), 'High-power impulse magnetron sputtering of Ti-Si-C thin films from a Ti3SiC2 compound target', Thin Solid Films 515(4), 1731-1736.

[3] Bandorf, R.; Vergöhl, M.; Giesel, P.; Wallendorf, T. & Mark, G. (2007), Investigation of HPPMS titania thin films prepared by unipolar, DC-superimposed and bipolar sputtering, in 'Society of Vacuum Coaters, 49th Annual Technical Conference Proceedings', 160-162.

[4] J.M. Schneider, W.D. Sproul, R.W.J. Chia, M.S. Wong and A. Matthews, Very-high-rate reactive sputtering of alumina hard coatings, Surf. Coat. Technol. 96 (1997), pp. 262–266.

[5] J.M. Schneider, W.D. Sproul, A.A. Voevodin and A. Matthews, Crystalline alumina deposited at low temperatures by ionized magnetron sputtering, J. Vac. Sci. Technol. A 15 (1997), pp. 1084–1088.

[6] Sarakinos, K.; Alami, J. & Konstantinidis, S. (2010), 'High power pulsed magnetron sputtering: A review on scientific and engineering state of the art', Surface and Coatings Technology 204(11), 1661 - 1684.

[7] Chistyakov, R. & Abraham, B. (2009), 'Advanced Pulsed DC Technology for Material Processing Applications', SVC Bulletin Spring 2009, 46-49.

[8] Chistyakov, R.; Abraham, B.; Sproul, W.; Moore, J. & Lin, J. (2007), 'Modulated Pulse Power Technology and Deposition for Protective and Tribological Coatings', Society of Vacuum Coaters 50th Annual Technical Conference Proceedings, 139-143.

[9] Chistyakov, R. & Abraham, B. (2006), 'Advances in High Power Pulse Reactive Magnetron Sputtering', Society of Vacuum Coaters 49th Annual Technical Conference Proceedings, 88-91.

[10] Bartzsch, H.; Glöß, D.; Böcher, B.; Frach, P.; Goedicke, K. (2003), 'Properties of SiO_2 and Al_2O_3 films for electrical insulation applicytions deposited by reactive pulse magnetron sputtering', Surface and Coatings Technology 174-175, 774-778.

Reactive HIPIMS with auxiliary Al electrode for ZnO:Al thin film deposition

Vasile TIRON, Claudiu COSTIN, Lucel SIRGHI, Gheorghe POPA

Alexandru Ioan Cuza University, Faculty of Physics, Iasi, 700506, Romania

Abstract. In this paper a new technique is proposed for precise doping control of ZnO:Al thin films deposited in reactive High Power Impulse Magnetron Sputtering (HIPIMS). An auxiliary aluminum electrode was added to a reactive Ar/O_2 pulsed magnetron with planar Zn target in order to obtain a controlled doping of ZnO films. Al neutral density in gas phase has been controlled by the discharge current and the biasing voltage on the auxiliary electrode (which influence the ion bombardment of the electrode) and measured by laser resonant absorption spectroscopy. The fraction of Al dopant in the deposited films was estimated by X-ray Photoelectron Spectroscopy (XPS) measurements. The goal of this work was to correlate Al density measured in the gas phase with Al concentration in the deposited films. It was also investigated the effect of the aluminum concentration on the structural, electrical and optical properties of ZnO:Al thin films deposited by HIPIMS. The internal microstructure and chemical composition of the deposited films was examined by X-ray difractometry (XRD) and X-ray Photoelectron Spectroscopy (XPS). The optical properties of the deposited films were studied by UV/VIS and photoluminescence spectroscopy.

Keywords: reactive HIPIMS, auxiliary electrode, aluminum doped zinc oxide

1. Introduction

For most optoelectronic devices, such as flat panel displays and solar cells, it is essential to use a transparent electrode, usually consisting of a thin film of transparent conducting oxide. ZnO:Al is a potential candidate for transparent electrodes in optoelectronic device due to the high transmittance in visible region and low resistivity [1]. Doping the ZnO with aluminum atoms rends the material conductor with preservation of its optical properties, which make this material a good candidate for synthesis of transparent conductive thin films.

A variety of methods for ZnO:Al thin film synthesis have been reported in the literature, e.g. chemical vapor deposition [2], molecular beam epitaxy [3], pulsed-laser deposition [4,5], RF and DC magnetron sputtering [6-8] and High Power Impulse Magnetron Sputtering (HIPIMS) [9]. Among these growth techniques, HIPIMS technique seems to be very promising as a physical vapor deposition method for surface pre-treatment and deposition of extremely dense, very smooth and highly adherent thin films in vacuum [10]. These serve as protective, functional coatings in a wide range of applications including hard, wear-resistant coatings, low friction coatings, corrosion-resistant coatings, decorative coatings and coatings with specific optical or electrical properties.

For doping control of ZnO films with aluminum there are used either co-sputtering of the two targets (Zn and Al or ZnO and Al_2O_3) [11] or sputtering of a composite target (Zn/Al or ZnO/Al_2O_3) [12]. In this paper, in order to control doping of ZnO with Al, an auxiliary aluminum electrode was

-----

Corresponding author.
E-mail address: vasile.tiron@uaic.ro (V Tiron)

added to a reactive Ar/O_2 pulsed magnetron with planar Zn target. Al concentration in the deposited film has been controlled by the magnetron discharge current and the biasing voltage on the auxiliary electrode.

2. Experimental setup

In order to obtain ZnO:Al conductor films, an auxiliary aluminum electrode, placed in the negative light of a magnetron discharge, was used to control the doping of ZnO films with aluminum atoms (figure 1). The vacuum chamber was evacuated to a base pressure of 10^{-6}Torr with a Dry Schroll rotary and turbomolecular pumps, and then the sputtering gas mixture, O_2 and Ar (99.99% purity for both cases), was introduced directly in the deposition chamber. The partial pressures of O_2 and Ar were regulated by mass flow controllers and adjusted at 1% ratio before the ignition of the discharge. During the film deposition, the total working pressure was maintained at 50 mTorr. All the ZnO:Al films were deposited for 60 min for the same magnetron discharge parameters (repetition frequency: 1 kHz, pulse width: 6 μs, target voltage on pulse: -900 V and target current on pulse: 15 A.

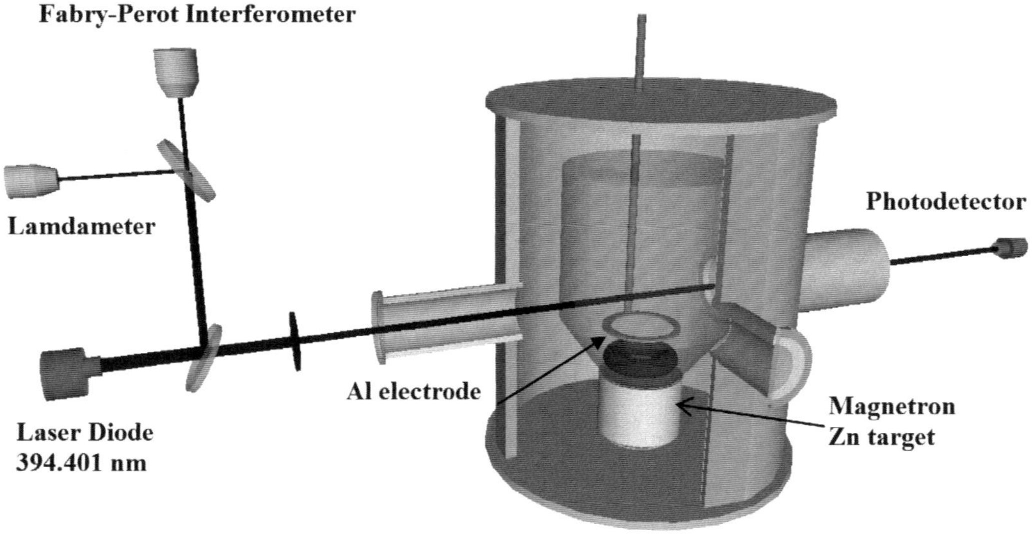

Figure 1. Experimental Setup

The pulsed magnetron discharge operates in a preionised pulsed regime with lower breakdown delay, high ion-to-neutral flux ratio at the substrate and high stability. By proper preionisation is possible to reduce the breakdown delay to the acceptable range to work in very short pulse regime (1-10μs) so that electric arcs are avoided [13]. Thus, stable operation are obtained at high frequency pulse repetition rate (~ 1 kHz) and high current densities (1-20 A/cm2). In figure 2 are illustrate the time evolutions of the discharge voltage and current during the pulse.

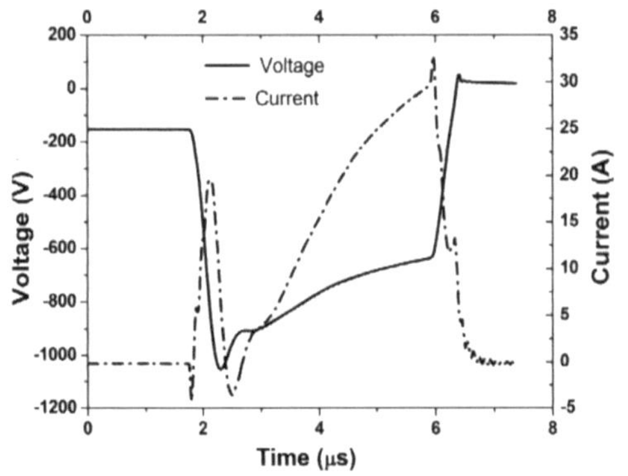

Figure 2. Time evolutions of the discharge voltage and current during the pulse; (gas pressure = 50 mTorr, O_2/Ar = 10%, frequency = 1 kHz, pulse width = 4 μs).

The aluminum electrode was placed in the negative glow of magnetron discharge and a negative DC voltage was applied in order to control the sputtering rate of aluminum atoms. To determine the concentration of aluminum atoms in the vapor phase and its spatial distribution, a tunable diode laser absorption spectroscopy method was used. A Toptica DL 100 diode laser system with an emitted radiation wavelength about 394.401 nm, corresponding to the aluminum transition from the ground state ($3s^23p$) to the excited state ($3s^24s$), was used. Laser absorption spectroscopy allows a real time monitoring of aluminum atoms density in the sputtering system [14].

3. Results and discussions

The aluminum atom concentration in gas phase has been controlled by the electric biasing potential of the electrode (which affected the ion bombardment of the electrode surface) and measured by laser resonant absorption spectroscopy. The obtained aluminum atom concentration in the deposited films was estimated by X-ray Photoelectron Spectroscopy (XPS) measurements.

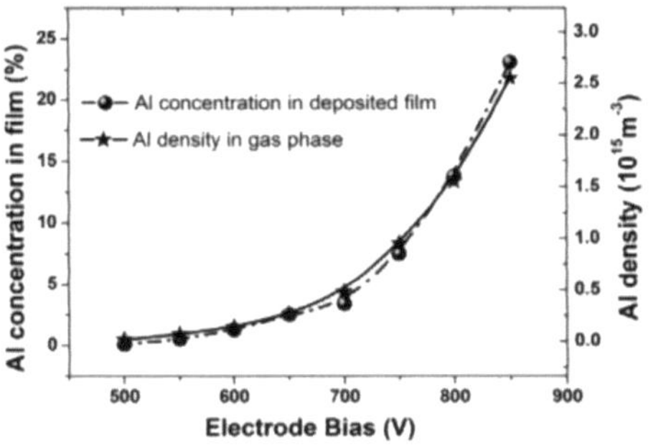

Figure 3. Aluminum atom concentration measured in gas phase and aluminum atom concentration in the deposited films; (gas pressure = 50 mTorr, O_2/Ar = 1%, pulse repetition frequency = 1 kHz, pulse width = 6 μs).

In figure 3 are plotted the aluminum atom concentration measured in gas phase and aluminum atom concentration in the deposited films as functions of electrode bias. In the both case the aluminum concentration increase rapidly with the electrode bias. Although the Al sputtering rate must be linearly dependent of the discharge power, in this case, the deviation is attributed to reactive operation mode of the auxiliary electrode discharge. Increasing the bias voltage the compound (Al oxide) is removed from electrode surface and Al sputtering rate increases. The optical transmission was measured by a UV-VIS spectrophotometer from 190 nm to 1100 nm. Figure 4a gives the spectra for all the ZnO:Al deposited films.

The fluctuation in the spectra is principally due to the interference effect owing to the reflection at interfaces. Sharp fundamental absorption edges are observed in all the spectra corresponding to the ZnO:Al films. These films have a high transmittance in visible regions and a high absorption in UV regions. As the Al content is increased the fundamental absorption edges become much gentler, with the reduction of the transmittance in visible regions and the absorption in UV regions, which is mainly attributed to the degradation of crystal quality. The thickness of deposited films was evaluated using the interference fringes from optical transmission spectra. The ZnO:Al thin films were deposited on fused quartz wafers substrates. The properties of deposited films are shown in table 1.

Table 1. Conditions for deposition and ZnO:Al films properties.

Sample	1	2	3	4	5	6
Bias Electrode (V)	500	550	650	700	750	800
Electrode Current (mA)	12	17	20	24	28	32
Thickness (nm)	720	740	730	820	1140	945
Band gap energy (eV)	3.46	3.56	3.28	3.30	3.31	3.56
Al concentration (%)	0.1	0.5	2.5	3.4	7.5	13.8

The photoluminescence (PL) measurements were performed at room temperature using a 340 nm UV Xenon lamp. The PL increases with the Al content in the deposited film.

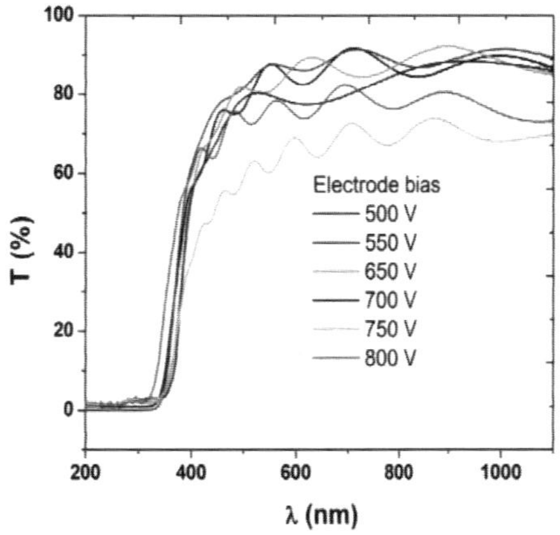

Figure 4a. The transmission spectra of ZnO:Al thin films deposited at different Al electrode biasing voltage.

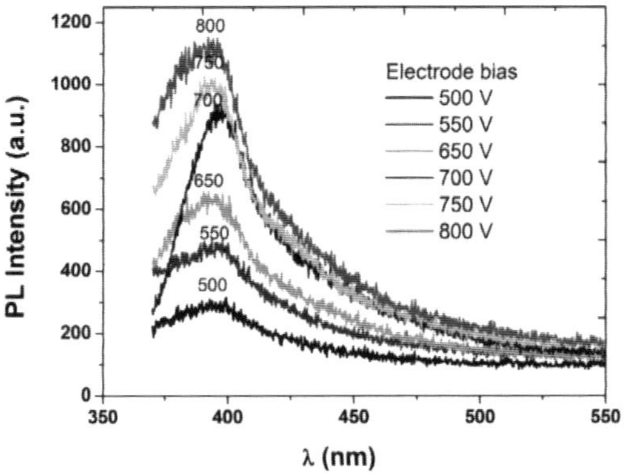

Figure 4b. The PL spectra (b) of ZnO:Al thin films deposited at different Al electrode biasing voltage.

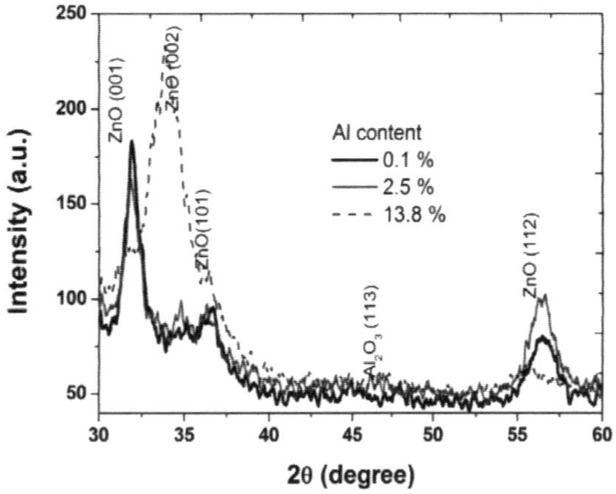

Figure 5. XRD profiles of ZnO:Al films prepared with the Al contents of 0.1, 2.5 and13.8 at.%.

The structural properties of the films were investigated by X-ray diffraction (XRD). When the Al content is small there is no preferential orientation in the film and no diffraction peak from other phase is detected, so the ZnO:Al film is a single-phase wurtzite ZnO structure. Increasing the Al content up to 13.8 at. %, the ZnO (002) diffraction peak becomes very high and wide suggesting that this film has *c*-axis preferential orientation and crystallites are very small (few nm). A new diffraction peak located at $2\theta = 47.0°$ is observed, which corresponds to the (113) plane of Al_2O_3, although its intensity is much smaller in the XRD pattern.

This is indicative of the appearance of the second phase. The ZnO:Al film in this case is a two-phase mixture consisting of ZnO hexagonal and Al_2O_3 rhombohedral structures.

4. Conclusions

A new technique for precise doping control of ZnO:Al thin films deposited in reactive High Power Impulse Magnetron Sputtering (HIPIMS) with auxiliary aluminum electrode was developed. Al concentration in ZnO:Al deposited films was tuned from 0.1% to 23% with preservation of its optical properties. There is a strong correlation between aluminum atom concentration measured in gas phase by tunable laser absorption spectroscopy and aluminum atom concentration in the deposited films measured by X-ray Photoelectron Spectroscopy (XPS). It was found that the obtained aluminum concentration in the deposited films was strongly dependent on the auxiliary electrode bias and magnetron discharge parameters. The photoluminescence response of ZnO:Al deposited films increases by increasing the Al doping concentration.

Acknowledgements

This work was supported by Romanian Ministry of Education, Research, Youth and Sport under grant POSDRU/89/1.5/S/49944.

References

[1] Ozgur U *et al* 2005 *J. Appl. Phys.* **98** 041301
[2] Lu J G and Fujita S 2007 *JOURNAL OF APPLIED PHYSICS* **101** 083705
[3] Triboulet R and PerrièreJ 2003 *Progress in Crystal Growth and Characterization of Materials* **47** Issues 2-3 65-138
[4] Kumar M *et al* 2005 *Thin Solid Films* **484** 174
[5] Agura H *et al Thin Solid Films* **445** (2003) 263–267
[6] Ruske F *et al* 2005 *Surface and Coatings Technology* **200** 236-240
[7] Liu W *et al* 2007 *Thin Solid Films* **515** 3057–3060
[8] Kluth O *et al* 2006 *Thin Solid Films* **502** 311 – 316
[9] Sittinger V *et al* 2008 *Thin Solid Films* **516** 5847
[10] Ehiasarian A 2010 *Pure Appl. Chem.* **82** No. 6 1247–1258
[11] Chung Y *et al* 2005 *Surface & Coatings Technology* **200** 936
[12] Cebulla R *et al* 1998 *J. Appl. Phys.* **83** No. 2 1087-1095
[13] Ganciu M *et al* 2005 *Journal of Optoelectronics and Advanced Materials* **7** 2481
[14] Vitelaru C *et al* 2010 *J. Phys. D: Appl. Phys.* **42** 124013

Microstructure, Oxidation and Tribological Properties of TiAlCN/VCN Coatings Deposited by Reactive HIPIMS

P Eh Hovsepian[1], G Kamath[1], A P Ehiasarian[1], R Haasch[2], I Petrov[2]

[1)] Sheffield Hallam University, Howard Street, Sheffield, S1 1WB, UK
[2)] University of Illinois at Urbana-Champaign, 104 South Goodwin Avenue, Urbana, Illinois 61801, USA

Abstract. 2.5 μm thick TiAlCN/VCN coatings were deposited by reactive HIPIMS process. XTEM showed gradual evolution of the structure of the coating with thickness. The initial structure is nanoscale multilayer with sharp interlayer interfaces. This transforms to nanocomposite of TiAlCN and VCN nanocrystalline grains surrounded by C-rich tissue phase and finally changes to an amorphous carbon rich Me-C phase. Energy-resolved mass spectrometry revealed that HIPIMS plasma is a factor of 10 richer in C^{1+} ions, and therefore more reactive, as compared to the plasma generated by standard magnetron discharge at the same conditions. The peculiar structure evolution in HIPIMS is is attributed to target poisoning effect.

Key words: High Power Impulse Magnetron Sputtering, nanoscale multilayer, reactive sputtering.

1. Introduction

Metal-carbon containing nanocomposite and nanolaminate structures have increased the interest of tool coating/machining industries, due to their inertness with Al-base alloys and low friction adaptability of carbon [1]. However, due to poor toughness and adhesion, state-of-the-art DLC coatings tend to fail prematurely when extreme thermal and mechanical loads are imposed during machining [2, 3]. Recently nanostructured TiAlCN/VCN multilayer coatings deposited by the combined HIPIMS/ UBM have shown considerable assurance to protect tools at elevated temperature applications [3, 4]. These coatings have revealed the ability to adapt self lubrication during sliding at elevated temperatures, (700 °C) by forming variety of Magnèli phase oxides [5]. We have previously reported TiAlCN/VCN multilayer coating deposited by mixed HIPIMS-UBM technology, which has shown very smooth, denser microstructure, with a very high adhesion to the substrate with enhanced mechanical and tribological properties at elevated temperatures of 650 °C [6]. In the present work, we report the microstructure and functional properties of TiAlCN/VCN nanoscale coating deposited by pure HIPIMS technology.

2. Reactive deposition by HIPIMS

The TiAlCN/VCN nanoscale coating was deposited onto bright annealed 304 stainless steel coupons 25 x 25 x 0.2 mm for structure analyses and surface characterisation. The coatings were deposited in an industrial sized coating machine (Hauzer Techno Coating, The Netherlands) enabled with HIPIMS technology at Sheffield Hallam University in UK using power supplies manufactured by Huttinger Electronic Sp. Z o.o. The coating machine comprises four rectangular cathodes (cathode size: 600mm x 200 mm). For this study, two opposing magnetrons furnished with TiAl (50:50 at%) and V

Corresponding author.
E-mail address: p.hovsepian@shu.ac.uk (P Eh Hovsepian)

targets, were operated in HIPIMS mode where as the other two magnetrons were shielded. Prior to deposition, the substrate surface was bombarded by V^+ + Ar^+ ions generated from a HIPIMS discharge sustained on the V target in Ar atmosphere using high bias voltage. To maintain constant voltage during the ion bombardment a HIPIMS dedicated power supply manufactured by Huttinger Electronic Sp. Z o.o. was used [7]. Detailed explanation of the surface pretreatment step was given elsewhere [8, 9]. In the next step of the process, an 800 nm thick TiAlN base layer was deposited. In the final stage of the coating deposition was carried out in a mixed $Ar+N_2+CH_4$ reactive atmosphere at 450 °C to produce TiAlCN/VCN.

3. Plasma compositional analysis

Time-averaged ion composition was determined by energy-resolved mass spectrometer PSM003 (Hiden Analytical Ltd.). The results for the deposition of the TiAlCN layer of the TiAlCN/VCN using HIPIMS show that the metal ion content was about 12% with Al^{1+} =9%, Ti^{1+} =2%, and C^{1+} =1%. In UBM mode, we detected factor 4 lower metal ion content of ~3% distributed among: Al^{1+} = 2.8%, Ti^{1+} = 0.23% and C^{1+} =0.13%. The ion flux in both processes was dominated by gas. The ratio of metal-ion to gas- ion concentration during HIPIMS deposition was 1:7. This ratio is significantly lower in UBM deposition of 1:32. Furthermore a high amounts up to 1% of free carbon were detected in the HIPIMS process compared to 0.13% in the UBM process.

4. Coating structure characterisation by GAXRD

In the present work, the GA measurements were performed at incidence angles of 1°, 2°, 5° and 10° the diffraction patterns are not shown here. The depths of penetration, xe which were averaged between 2θ values of 20-110° and assuming intensity attenuation1/e were as follows: 0.360 μm, 0.720 μm, 1.798 μm and 3.583 μm at the incidence angles used. Depending on the incidence angle the shape of the diffraction peaks changes from broader and lower intensity peaks at small angles, (1°, 2°) to sharper and higher intensity peaks as well as additional reflection from {311} and {222}plains at 5° and 10°. This is confirmed by the FWHM measurements where for the (111) reflection the values change from 2.77° at 1° to 1.55° at 10°. GAXRD characterisation indicate that the structure of the coating gradually changes with thickness from larger grain crystalline at the base to random orientation smaller grain nanocrystalline to almost X-ray amorphous structure on the top.

5. Structure evolution of TiAlCN/VCN nanoscale coating deposited by HIPIMS

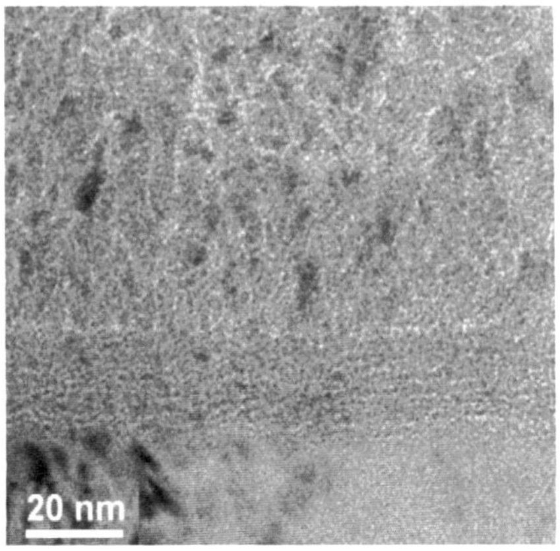

Figure 1a. BF TEM cross-section of TiAlCN/VCN nanoscale coating deposited by reactive HIPIMS deposition: Lower magnification image.

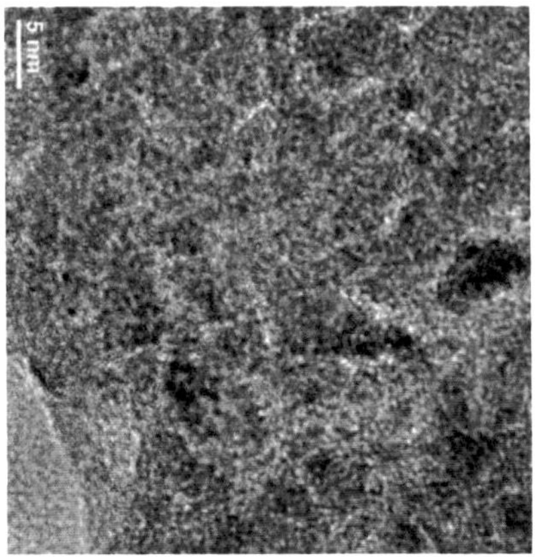

Figure 1b. BF TEM cross-section of TiAlCN/VCN nanoscale coating deposited by reactive HIPIMS deposition: High resolution image.

Cross-sectional TEM imaging revealed interesting details of the fine structure of the HIPIMS deposited TiAlCN/VCN, Figure 1. A nanoscale multilayer structure with bi-layer thickness of 5-8 nm and total thickness of 30 nm can be observed right above the TiAlN base layer. Careful observation reveals a white lateral phase segregated at the interfaces of the individual layers. This structure is typical for TiAlCN/VCN coatings deposited by sputtering and its growth mechanism was described previously in [10]. The structure which evolves further with coating growth is one of a typical nanocomposite consisting of dark contrast grains (most probably MeCN where Me is Ti-Al-V) surrounded by carbon-based phase forming a white contrast tissue phase. The largest grains with diameter in the range of 5-10nm are formed right after the band with nanoscale multilayer structure. Figure 1b is a High Resolution TEM micrograph showing the nanocrystalline grains and the surrounding tissue phase structure. The grain size reduces gradually with thickness at the expense of the white phase and no grains can be seen on the very top region of the coating. Point EDX analyses were carried out in different zones of the coating namely base layer and nanoscale multilayer, bulk of the coating and top of the coating, Fig. 2 a, b, c.

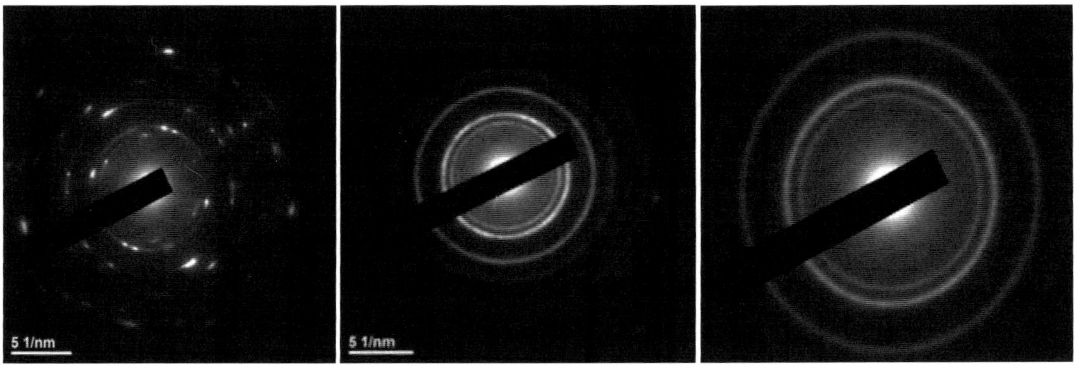

Figure 2. EDX patterns taken from various coating zones: a) TiAlN base layer/ nanoscale multilayer zone, b) nanocomposite zone, c) top of the coating.

The analyses showed that a gradual transformation in the structure with thickness of the coating takes place changing from crystalline at the base to nanocrystalline in the bulk and finally random orientation small grain nanocrystalline to almost X-ray amorphous structure on the top thus confirming the GAXRD findings and providing further information for better understanding and interpretation of the XTEM results. AES depth profile (not shown here) revealed that the C content in the film gradually increased with the film thickness from 25at% at the TiAlN base layer-TiAlCN/VCN coating interface to 70% at the very top of the coating. This unique structure and obviously compositional evolution obtained in one deposition process is believed to be due to the higher content of C ions in the plasma and the gradual progression of the target poisoning effect which takes place in the reactive (carbon-nitrogen containing atmosphere) HIPIMS.

6. Oxidation resistance by Thermo Gravimetric analysis

The oxidation resistance of TiAlCN/VCN coating deposited by reactive HIPIMS technique was studied by TG Analyser from SETARAM instrumentation and compared to the performance of conventional UBM and mixed HIPIMS+UBM deposited coatings. All measurements wear performed with linear temperature ramp at 1°C/min from room temperature to 1000 ^{o}C. It was revealed that the onset of rapid oxidation for the HIPIMS coating was shifted to higher temperatures of about ≈ 810 ^{o}C as compared to the UBM (≈ 550 ^{o}C) and HIPIMS+UBM, (≈ 780 ^{o}C) cases. Furthermore, the weight gain due to oxidation was lower by factor of two when compared to UBM deposited TiAlCN/VCN coating which demonstrates the effect of the coating densification due to utilisation of HIPIMS.

7. Friction and wear behaviour of the coating at room temperatures

Coating friction and wear behaviour was investigated by pin-on-disc tests using Al_2O_3 counterpart. The friction curve at room temperature depicted clearly two friction zones. Immediately after the "running" stage the friction curve changes its slope indicating the beginning of the first lower coefficient of friction zone where the mean value of μ=0.37. After that the friction gradually increases to reach a "steady state" where higher coefficient of friction values (μ=0.48) were recorded. In an attempt to "depth profile" the friction and wear behaviour of the HIPIMS deposited TiAlCN/VCN coatings a number of friction tests were carried out with stepwise increase of the sliding distance. In this experiment it was found that the wear coefficient reduces by one order of magnitude with the wear depth, with initial value of $K_c = 3.8 \times 10^{-15}$ $m^3 N^{-1} m^{-1}$ after 500 laps to $K_c = 4.4 \times 10^{-16}$ $m^3 N^{-1} m^{-1}$ after 10000 laps. In comparison, UBM deposited TiAlCN/VCN coatings tested under similar conditions show one order of magnitude higher wear coefficient of 4.1×10^{-15} $m^3 N^{-1} m^{-1}$ [6]. This peculiar wear behaviour is believed to result from the graded coating phase composition and micro structures. Coefficient of friction values of μ=0.37 as measured in the first friction zone at the top of the coating are typical for Me-DLC coatings. This corresponds well to the presence of Me-C phase and XRD amorphous structure of the coating. Low friction in this zone is achieved due to the presence of C, which acts as a solid lubricant. In the nanocomposite region of the coating, the coefficient of friction increases as compared to the C-rich top layers however the wear resistance improves due to the higher hardness of the nanocrystalline structures. In this zone the friction and wear behaviour are determined by the efficiency of the solid lubrication process due to the presence of C and formation of V_2O_5 Magnèli phases during sliding [6].

8. Conclusions

- GAXRD and XTEM analyses showed that during HIPIMS deposition of TiAlCN/VCN a gradual transformation in the structure with thickness of the coating takes place. The structure evolves from crystalline nanoscale multilayer of TiAlCN/VCN at the base to nanocomposite structure comprising nanocrystalline TiAlVCN surrounded by C-based tissue phase in the bulk and finally to X-ray amorphous structure of Me-C on the top. This unique structure and compositional evolution obtained in one deposition process is believed to be due to the higher content of C ions in the plasma and the kinetics of the target poisoning effect during deposition by HIPIMS in reactive (carbon-nitrogen) atmosphere.
- TG analyses revealed that the onset of rapid oxidation for HIPIMS deposited TiAlCN/VCN coatings was shifted to higher temperatures, (to about 810 ^oC) as compared to the UBM and HIPIMS+UBM cases. The weight gain due to oxidation of the HIPIMS coating was factor of two lower than that for the UBM coating due to the higher density of the HIPIMS coating.
- "Depth profiling" the friction and wear behaviour of the HIPIMS deposited TiAlCN/VCN coatings revealed that immediately after the "running" stage, a low friction coefficient (μ = 0.37) zone exists. After that the friction gradually increases to reach a "steady state" with coefficient of friction of μ = 0.48. The wear coefficient reduces by one order of magnitude with wear depth, with initial value of $K_c = 3.8 \times 10^{-15}$ $m^3 N^{-1} m^{-1}$ after 500 laps to $K_c = 4.4 \times 10^{-16}$ $m^3 N^{-1} m^{-1}$ after 10000 laps. In comparison, UBM deposited TiAlCN/VCN coatings tested under similar conditions show an order of magnitude higher wear coefficient of 4.1×10^{-15} $m^3N^{-1}m^{-1}$. This peculiar wear behaviour is believed to result from the combined effects of the graded microstructure and densification through the HIPIMS process.

IOP Conf. Series: Materials Science and Engineering **39** (2012) 012011 doi:10.1088/1757-899X/39/1/012011

References

[1] M.Lahres, P.Müller-Hummel, O. Doerfel, Surf. Coat. Technol. 91(1997)116-121.
[2] A. A. Voevodin, J.S. Zabinski, Compos. Sci. Technol. 65 (2005) 741.
[3] P. Eh. Hovsepian, A.Ph. Ehiasarian, A. Deeming, C. Schimpf, Vacuum, 82 (2008), 1312– 1317.
[4] P. Eh. Hovsepian, A. P. Ehiasarian, U. Rataysaki, Surface and Coatings Technology 203 (2009) 1237-1243.
[5] G. Kamath, A. P. Ehiasarian, P. Eh. Hovsepian, Surf. Coating and Technol, 205, 8-9, (2011) 2823-2829.
[6] G. Kamath, A. P. Ehiasarian, P. Eh. Hovsepian, IEEE Transaction on Plasma Science, 38, (11), 2010, 3062-3070.
[7] A. P. Ehiasarian, R. Tietema, P. Hovsepian, D. Doerwald, R. Bugyi, A. Klimczak, Pat appl. US2010/0025230A, Apr. 10.2007.
[8] A. P. Ehiasarian, P. Eh. Hovsepian, W.-D.Münz, US Pat. 7081186B2, 2005, EP 1 260, 603 A2, DE 10124749, 21.05. 2001.
[9] A. P. Ehiasarian, J. G. Wen and I. Petrov, Interface microstructure engineering by high power impulse magnetron sputtering for the enhancement of adhesion, J. Appl. Phys. 101 (5) (2007).
[10] P.Eh. Hovsepian, A.P. Ehiasarian, I. Petrov, Surf. Engin. 26 (2010) 610.

Plasma self-organisation in High Power Impulse Magnetron Sputtering Discharges

A P Ehiasarian [a], A Hecimovic [b], J Winter [b], T de los Arcos [b], R New [a], V Schulz-von der Gathen [b], M Böke [b],

[a] HIPIMS Technology Centre, Sheffield-Hallam University, UK
[b] Institute of Experimental Physics II, Ruhr-University Bochum, Germany

Abstract. We report for the first time on instabilities of High Power Impulse Magnetron Sputtering Plasmas (HIPIMS) which are likely to be of the generalised drift wave type. They are characterized by well defined regions of high and low plasma emissivity along the racetrack of the magnetron and causes periodic shifts in floating potential. The azimuthal mode number *m* depends on plasma current, plasma density, and gas pressure. The structures rotate in $\vec{E}\times\vec{B}$ direction at velocities of ~10 kms^{-1} and frequencies up to 200 kHz. Collisions with residual gas atoms slow the rotating wave down, whereas the elevated plasma density accelerates it. The instabilities result in intermittent modulated particle fluxes with significant magnitude transported away from the cathode.

1. Introduction

High power impulse magnetron sputtering (HIPIMS) is the latest innovation in physical vapour deposition technology [Kouznetsov]. It combines high voltage / high current impulse glow discharges with conventional magnetron cathodes to achieve highly ionised sputtered flux. Heat flux management at the cathode is achieved by pulsing the discharge at low duty cycle with typical durations of a few 100 μs and frequencies of a few 100 Hz. The ionisation degree can be controlled in a wide range by the peak power in the discharge. It allows the tailoring of a number of different properties of the films [Sarakinos].

The HIPIMS plasma discharge is often reported to be a homogeneous however, perfect homogeneity is seldom observed in discharges in general. For example, unmagnetized high power DC glow discharges show tendencies for self-organisation resulting in structures of modulated intensity at the cathode [Benilov]. The high discharge current density in HIPIMS plasmas of the order of Acm^{-2} and the presence of crossed electric and magnetic fields induce a strong azimuthal $\vec{E}\times\vec{B}$ drift [Rossnagel, Vetushka, Brenning].

Magnetrons are vulnerable to two-stream instabilities that occur due to large differences in fluxes of electrons and ions [Lundin, Brenning]. The strength of the magnetic fields close to the cathode surface is such that electron gyro radii are much smaller than the magnetic trap, whereas ion gyro radii are much larger. Therefore electrons are magnetized and confined whereas ion loss rates are high. The HIPIMS configuration is topologically is similar to that of Hall thrusters. Their instabilities have been analyzed and classified by Choueri [Choueri] pointing out several drift wave type instabilities.

Recent fast-frame photography has shown that indeed HIPIMS plasmas exhibit striations - faint bands with alternating low and high intensity - in a direction perpendicular to the $\vec{E}\times\vec{B}$ drift [Ehiasarian]. In this contribution we report the first observation of centimetre-size, well-defined bright

-----

Corresponding author.
E-mail address: ante.hecimovic@rub.de (A Hecimovic)

and dark regions of plasma arranged periodically along the full length of the racetrack of a circular HIPIMS magnetron. As will be shown, these features resemble the footprint of drift-wave lobes with well defined azimuthal periodicity, i.e. mode number *m*. They rotate with frequencies in the range of typically several 100 kHz and velocities of ~10 kms^{-1}.

2. Experimental

Experiments were conducted at Ruhr-University, Bochum in a UHV cylindrical vacuum chamber of 45 cm height and 35 cm diameter, pumped by a turbomolecular pumping system to a base pressure of 1x10^{-6} mbar. We employed a standard circular magnetron with a field created by a single cylindrical magnet acting as the inner pole and a monolithic ring magnet as the outer pole. The target diameter was 50 mm and the magnetic field strength on the target surface was 150 mT (50 mT at 5 mm from surface).The inner pole was south and the outer pole was north resulting in an $\vec{E}\times\vec{B}$ drift in anticlockwise direction. The magnetron has a slightly unbalanced configuration. A HIPIMS power supply type HMP 2/1 (Huettinger Electronic Sp. z o.o.) was used to drive the discharge. Typical voltage-current traces are shown in Figure 1.

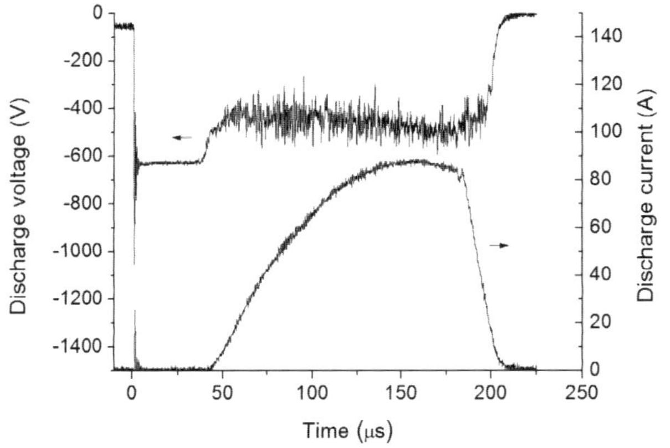

Figure 1. Typical waveforms of the discharge voltage and discharge current.

To study the discharge with high temporal and spatial resolution we observed the cathode surface face on through a glass view port with wavelength cutoff at 300 nm. Fast photography was realised using a single shot camera type PIMAX 1024 (Princeton Instruments) with image intensifier, equipped with a UV-Nikkor 105 mm f 1:4.5 objective. The camera exposure was synchronised with the discharge pulses and the CCD detector was gated for 150 ns through an electronic shutter. The gate width was chosen to be shorter than the characteristic motion time of the then observed structures, in particular exposures of 30, 70, 150, 300 ns revealed no change in the localization of the structures whereas longer exposures of 1000 and 2000 ns showed significant smearing of the image. The floating potential with respect to ground was measured using a single Langmuir probe with length of 5 mm and diameter of 50 μm, positioned above the magnetron at a distance of 19 mm and at a radial distance of 25 mm from the magnetron centre (Fig. 2).

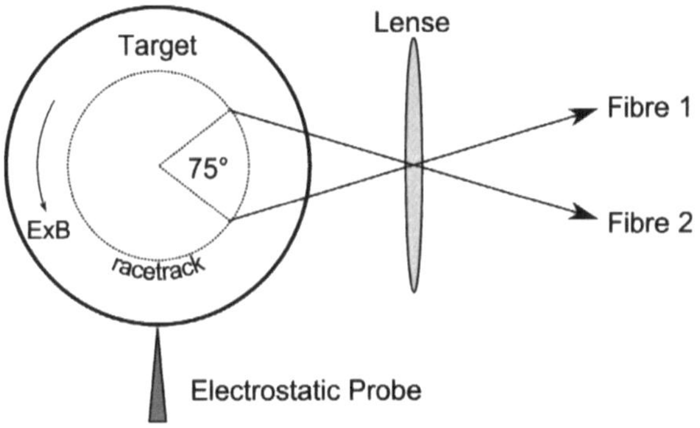

Figure 2. Unfolded view of the target with arrangement of Langmuir probe and local optical measurement. The actual light paths were in a plane perpendicular to the target.

The local intensity of total light from two well defined points on the target was imaged onto two photomultipliers and recorded independently (Figure 2). The optical foci were put at the position of the racetrack above the target surface at an angular distance of 75° and linear distance of 20 mm with respect to the estimated racetrack centreline. The radius of the racetrack was 15 mm and the length of the racetrack along its centreline was around 100 mm. The points were selected by projecting an image of the target onto two fibre optic cables positioned at the focal length of a single lens as shown in Figure 2. The light from each fibre was detected by separate photomultiplier tubes (PMT) type Hamamatsu R928 and Nucletron 1101 and fed to an oscilloscope (LeCroy Waverunner LT374M) through equal terminations (resistance of 1kΩ and capacitance of 16 pF) yielding a time constant of the recording system of ~20 ns. The speed of motion was calculated from the time shift and the distance between observed points. The frequency of rotation was obtained from Fast Fourier Transform (FFT) analysis of the signals.

3. Results and discussion

Figure 3 shows images of the race track (total light) emission at different peak current densities (J_d) and at different Argon pressure (P_{Ar}). At low currents the plasma is essentially homogeneously distributed around the race track (Figure 3a). At high currents (Figure 3b), there is one distinct region of low intensity where the signal drops to 50% of the maximum, as shown on Figure 3b. At high pressures (Figure 3c), we recognise four distinct structures with similar appearance and size.

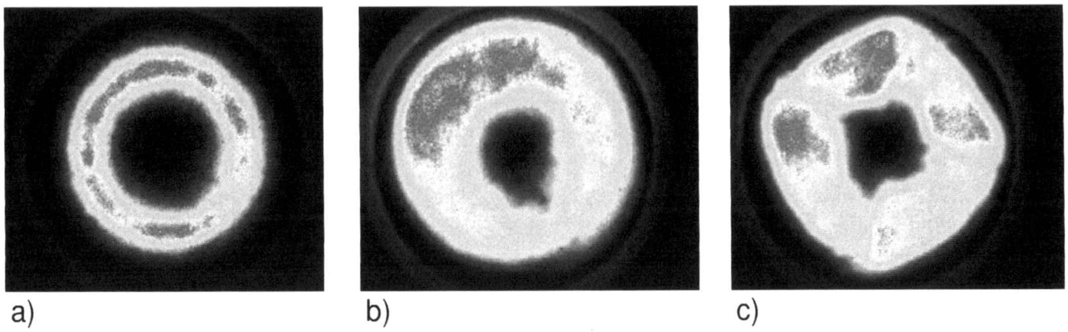

Figure 3. Influence of peak current density and Ar pressure on the discharge: a) $J_d = 0.75$ Acm^{-2} , P_{Ar} = 1.7 Pa; b) $J_d = 7.5$ Acm^{-2} , P_{Ar} = 1.7 Pa; and c) Jd =7.5 Acm^{-2}, P_{Ar} = 16.5 Pa.

Signals from the individual optical fibres and of the floating potential as a function of time are given in Figure 4.

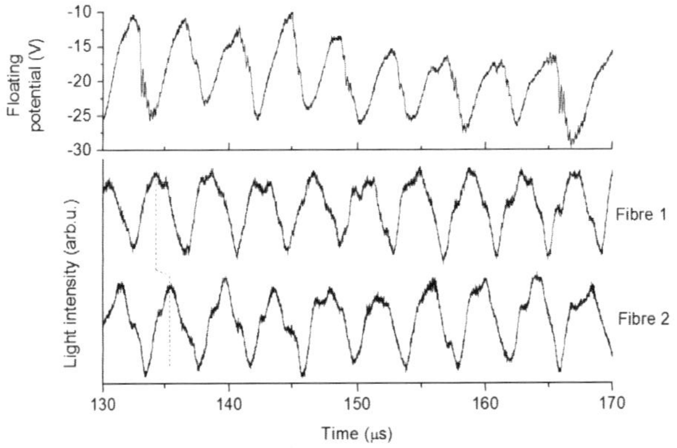

Figure 4. Signals from optical fibres and the floating potential measured by electrostatic probe at J_d = 7.5 Acm^{-2} , P_{Ar} = 1.7 Pa.

All signals have the same well-defined periodicity. Quantitative treatment by FFT (not shown) confirms a well-defined peak at a characteristic frequency with second and third order harmonics clearly present. This indicates that, for the observed time period and at the given parameters, the structures are stable in time and maintain a constant amplitude, frequency and wavelength. This is characteristic of e.g. saturated drift mode type instability.

The minima of the electrostatic probe signal can be interpreted as the increase in the electron density and therefore correlate with PMT maxima due to increase in the excitation and emission in the part of the plasma with elevated electron density.

Comparing the signals from the two optical fibres as shown in Figure 4 we observe a time shift of 1.3 μs, marked with dotted line. Using recognisable features and varying the angle between the two fibres we found that the phase shift is in positive direction indicating that the periodic structures move anticlockwise along the track following the direction of $\vec{E}\times\vec{B}$ drift expected for this cathode. The PMT signals in conjunction with the camera images show unambiguously that there is a wave of particles propagating through the plasma in the direction of the $\vec{E}\times\vec{B}$ drift responsible for excitation of atoms.

The experimental values for frequency and time shift obtained from PMT measurements were used to calculate velocity, wavelength and expected number of lobes within the racetrack circumference in Table 1. The actual number of features are those observed in camera shots in Figure 5.

Table 1. Influence of gas pressure at peak current of 150 A (7.5 Acm^{-2}), on the parameters of the rotating structures. Wavelength was calculated by dividing velocity by frequency.

Pressure (Pa)	**Frequency (kHz)**	**Time shift (μs)**	**Velocity (kms^{-1})**	**Wavelength (mm)**	**Number of Waves on Target**	**No. of Observed Features**
1.69	188	1.39	15.0	79.7	1.25	1
10	368	1.57	13.3	36.1	2.8	3
16.3	445	1.85	11.3	25.3	4	4

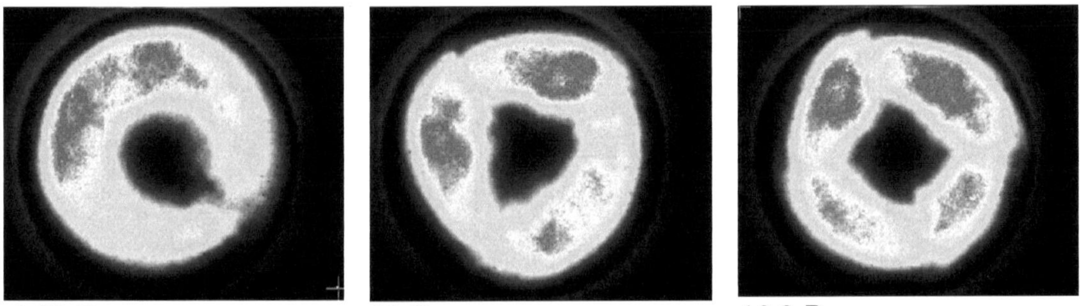

Figure 5. Camera shots at different pressure corresponding to calculations in Table 1.

The frequencies of the waves are between 100 and 200 kHz, and are considerably smaller than the lower hybrid frequency which is of the order of 2 MHz [Lundin]. We cannot exclude at this stage that there may be faint high frequency components superimposed on the wave patterns, which could be indicative of two stream instabilities as discussed by Bohlmark [Bohlmark]and Lundin [Lundin].

The exact nature of the observed instability is still unclear. The magnetic topology in these experiments is more complex than in e.g. simple drift wave experiments in cylindrical plasmas confined by co-axial homogeneous magnetic fields.

Furthermore HIPIMS is a transient discharge in which the current density and its influence on the actual magnetic field varies strongly. The measurements with magnetic pick-up coils (not shown here) indicate [Bohlmark, Barwe] that the current-induced magnetic field far away from the magnetron may be of the same order as the vacuum magnetic field.

Another complication is the mirror confinement of the electrons. This may introduce a two-stream instability, which should occur, however, at a much higher frequency than observed for our driven wave. A self consistent PIC calculation allowing for the real magnetic configuration may be needed to resolve the puzzle. Following Choueri [Choueri] we assume at this stage that our observed waves are essentially of the generalised drift wave type.

4. Conclusions

Our observations clearly demonstrate that HIPIMS plasma is highly complex and inhomogeneous form of the plasma discharge. The phenomena observed and presented in this paper are most probably drift wave instabilities. Under certain experimental conditions the instabilities saturate and result in structures with various azimuthal periodicity (m-number) rotating at frequencies between 100 kHz and 200 kHz. The rotation speed is slowed down by collisions with the residual gas and is dependent on the conductivity of the medium. The instabilities result not only in intermittent modulated particle fluxes away from the cathode but most likely also to highly localized transient heat loads on the cathode surface, which may be of significant technological importance. The nature and properties of these particle fluxes still have to be investigated. It may be speculated that the observed modulated floating potential may be due to particle beam (beacons) emitted from the cathode area outwards.

Acknowledgement

This work was supported by DFG within Project A5 of SFB-TR87 and by the Research Department "Plasmas with complex Interactions" of Ruhr-University Bochum.

5. References

Barwe. B, Investigation of High Power Pulsed Magnetron Sputter Plasmas (HPPMS) by spectroscopic and electric measurements, Diploma work, Ruhr University Bochum 2010

Benilov MS, Phys. Rev. E - Statistical, Nonlinear, and Soft Matter Physics. 2008; 77(3):036408.

Bohlmark J, Helmersson U, VanZeeland M, Axnäs I, Alami J, Brenning N. Measurement of the magnetic field change in a pulsed high current magnetron discharge. Plasma Sources Sci. Technol. 2004;13(4):654-661.

Brenning N, Axnäs I, Raadu MA, Lundin D, Helmerson U. A bulk plasma model for dc and HiPIMS magnetrons. Plasma Sources Sci. Technol. 2008;17(4):045009.

Choueri E. Y., Phys. Plasmas 8 (2001), 1411

Ehiasarian AP, in Plasma Surface Engineering Research and its Practical Applications, edited by R. Wei (Research Signpost, Kerala, India, 2008), pp. 35-86.

Kouznetsov V, Macak K, Schneider JM, Helmersson U, Petrov I. A novel pulsed magnetron sputter technique utilizing very high target power densities. Surface and Coatings Technology. 1999 15;122(2):290-3.

Lundin D, Larsson P, Wallin E, Lattemann M, Brenning N, Helmersson U. Cross-field ion transport during high power impulse magnetron sputtering. Plasma Sources Sci. Technol. 2008;17(3):035021.

Rossnagel SM , and Kaufman HR, J. Vac. Sci. Technol. 1987; A5(1):88.

Sarakinos K, Alami J., Konstantinidis S., Surface and Coatings Technology, 204 (2010) 1661-1684

Vetushka A, Bradley JW. The current-density distribution in a pulsed dc magnetron deposition discharge. J. Phys. D: Appl. Phys. 2007;40(7):2037-2044.

IOP Conf. Series: Materials Science and Engineering **39** (2012) 012013 doi:10.1088/1757-899X/39/1/012013

Time-resolved laser-induced fluorescence diagnostics in a HIPIMS discharge

Nikolay Britun[1], Maria Palmucci[1], Stephanos Konstantinidis[1], and Rony Snyders[1,2]

[1] Chimie des Interactions Plasma Surface (ChIPS), CIRMAP, Université de Mons, Place du Parc 20, B-7000 Mons, Belgium
[2] Materia Nova Research Center, 1 Avenue Copernic, B-7000 Mons, Belgium

Abstract. Laser-induced fluorescence (LIF) combined with resonant optical absorption spectroscopy (ROAS) were utilized for characterization of the Ar-Ti high-power impulse sputtering (HIPIMS) discharge at working pressures of 5 and 20 mTorr. The LIF measurements were performed during the plasma off-time. It was shown that the time-behavior of the measured densities depends strongly on the working pressure in the reactor as well as on the distance from the magnetron target. ROAS measurements indicates that the absolute ground state level density of Ti can reach $\approx 10^{11}$ cm^{-3} (at 20 mTorr, 10 μs pulse, and 12 kW of applied power per pulse). Generally, the obtained results indicate highly non-uniform spatial and temporal behavior of the sputtered species such as Ti and Ti^+, which correlate with the results previously obtained in the other magnetron discharges.

1. Introduction

There are many advantages which determine the usefulness of the optical plasma diagnostics methods for the sputtering discharges, and particularly for the high power impulse magnetron sputtering (HIPIMS). Among them: (i) being characterized optically the plasma remains unperturbed, (ii) the sputtered particles as well as the presented electromagnetic field can be an obstacle for intrusive diagnostics methods, (iii) many parameters such as the gas density, its temperature, etc. can be deduced based on the optical diagnostics, (iv) time-response of the optical methods is often lies in the ns range, which is excellent for the fast changing plasmas such as the HIPIMS discharge. Among the most popular plasma diagnostics methods are optical emission spectroscopy (OES), resonant optical absorption spectroscopy (ROAS), laser-induced fluorescence (LIF), etc. Each of them has its own advantages and drawbacks, e.g. LIF is recognized for its high spatial (<1mm) and temporal (~10ns) resolution [1], as well as for the narrow laser linewidth, especially when using the laser emission diodes [2]. LIF can be applied to analyze the non-excited gases, but at the same time it provides only relative density of species, and involves rather bulky and expensive equipment. ROAS gives an absolute density of species [3] in a discharge, it does not depend on the plasma excitation as well, however it has somewhat limited sensitivity and requires an additional light source. OES is very flexible and easily accessible diagnostics which is widely used for qualitative characterization of plasma; however it requires a substantial amount of the excited species in a discharge [4]. In the present work the results of LIF diagnostics is combined with another optical plasma characterization methods to study the HIPIMS discharge.

Corresponding author.
E-mail address: nikolay.britun@umons.ac.be (N Britun)

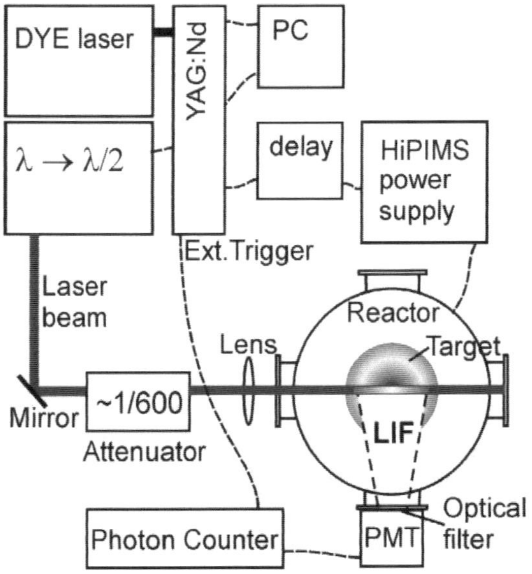

Figure 1. The schematic representation of the setup for the HIPIMS discharge diagnostics by laser-induced fluorescence.

2. Experimental

The HIPIMS discharge used in this work was operating at 1 kHz, and has 10 μs of the pulse duration. Two working pressures, namely 5 mTorr and 20 mTorr, were used in the experiments. The plasma characterization was performed during non-reactive (Ar) sputtering of a 10 cm diameter water cooled Ti target. The peak current of the HIPIMS plasma pulse was about 60 A, the voltage was about 500 V giving the average power per pulse of about 12 kW.

The schematics of the LIF experimental setup is presented in Fig.1. The INDI YAG:Nd laser was utilized for pumping the Sirah CobraStretch dye laser at the wavelength of 532 nm. The second harmonic laser beam generated by a BBO crystal and attenuated by a factor of about 600 was directed into the HIPIMS reactor, so the energy of the laser pulse in front of the target was about 5 μJ. Such a low value of the laser energy is important for the linear response of LIF measurements [5]. The cylindrical plasma volume excited by the laser parallel to the magnetron target with the length of about 10 cm and diameter less than 1 mm was studied by LIF. Three distances from the target surface were investigated in this work (15, 45, and 70mm). The LIF signals from Ti and Ti^+ were measured by a Hamamatsu R928 photomultiplier tube (PMT) after filtering by a band-pass optical filter with 10 nm spectral width. The spectral transitions used for the LIF measurements were as follow: Ti: $\lambda_{excitation}$ = 320.58 nm, $\lambda_{fluorescence}$ = 509.66 nm; Ti^+: $\lambda_{excitation}$ = 338.38 nm, $\lambda_{fluorescence}$ = 486.56 nm. By exciting the listed transitions, the relative density of the lowest ground state level (0 cm^{-1}) for both Ti and Ti^+ were determined. To measure the fluorescence signal coming from the excited volume of the discharge, the PMT was biased by a DC voltage of about 1.1 kV to reach the photon counting regime of operation. The SRS430 photon counter with 40 ns of time resolution was utilized for final accumulation of the LIF signal. For the absolute Ti density estimations by ROAS an Andor Shamrock750 spectrometer-monochromator coupled with an Andor iStar740 ICCD camera was utilized. The Cathodeon Ti hollow cathode lamp in the pulse mode was utilized as a reference source.

3. Results

3.1. Plasma lines broadening

In order to determine properly the species density, the additional measurement of the spectral profiles of the plasma lines (i.e. the line broadening) has to be performed. Nevertheless, in the pulsed discharges the lines broadening can change substantially as a function of time, which reflects the fact that these discharges are far from thermal equilibrium. Since the full width at half maximum (FWHM) of the exciting laser is normally a constant, the change of the plasma linewidth may affect the total density measured by LIF. It is especially the case if the overlapping between the laser and plasma lines is substantially different from one measurement to another. Our measurements performed for few typical discharge conditions reveal very prominent time-dependences for the plasma linewidths. The summary of these measurements after deconvolution with the laser line profile (0.8 pm at 320nm in our case [6]) are given in Fig. 2.

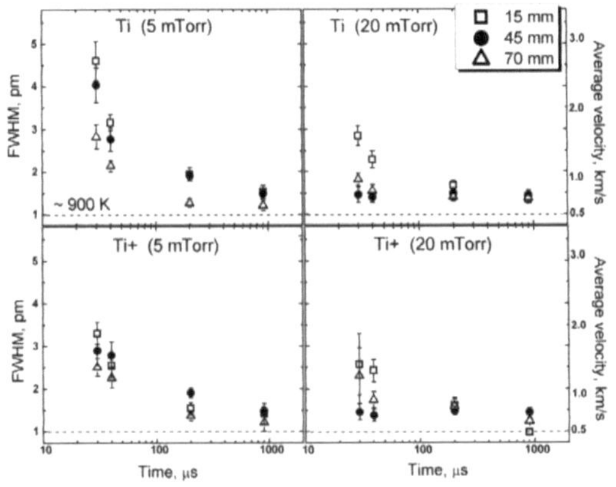

Figure 2. The time-resolved FWHMs of the Ti and Ti^+ plasma spectral lines after deconvoluiton with the laser profile measured by LIF at three different distances from the magnetron target.

As we can see, the FWHMs of Ti and Ti+ spectral lines decrease rapidly reflecting very fast thermalization of species during the plasma off-time. More careful analysis of the obtained broadening values shows a faster-than-an-exponential decay during the plasma off-time. The spectral shape of the measured lines is always close to the Lorentzian one. The rapid FWHM decay is more pronounced at low distance to the target, where thermalization degree is supposed to be low (empty squares in Fig.2). As one can expect, the FWHM are generally lower at higher Ar pressure, which is also due to the faster thermalization at this pressure. At the very end of the plasma off-time the measured FWHMs reach the values of about 1 pm, which correspond to the average Ti velocity parallel to the target of about 500 m/s, which roughly can be assigned to 900 K of the gas temperature. The temperature, however, cannot be directly calculated in this case, mainly due to the spectral shape of the measured lines is found to be close to Lorentzian (rather than to the Maxwellian distribution).

3.2. Density of species

Knowledge of plasma lines broadening in a definite moment of time (or the velocity distribution of the species in the discharge at this time), enables to calculate the total density of species by taking an integral of the total LIF signal measured at different laser wavelengths in the vicinity of the LIF transition for definite species [7]. In this way the time dependences for the relative density of the measured species (Ti, Ti^+) during the plasma off-time were determined as a function of time and spatial position. The results are presented in Fig.3, from where one can observe that at higher pressure the maximum density of sputtered Ti is about 5 times higher (be aware of the different scale for Ti density), than that at lower pressure (at about 50 μs after the plasma pulse).

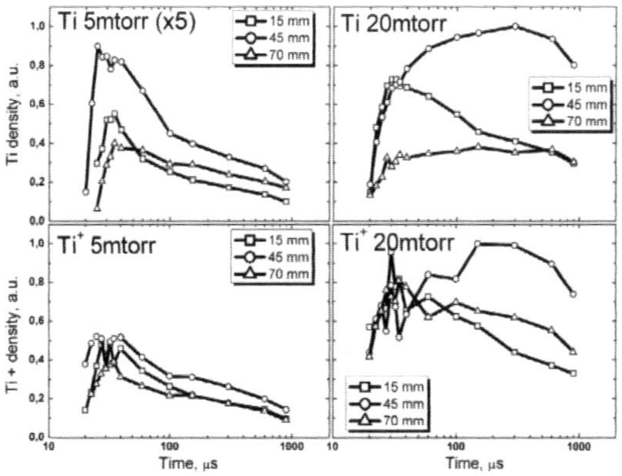

Figure 3. Normalized (separately for Ti and Ti^+) time-relolved relative density of Ti and Ti^+ in the HIPIMS reactor measured by LIF at three different distances from the magnetron target for typical discharge conditions studied in this work.

The ion density at the same time increases roughly twice. This can be explained by a stronger sputtering at higher pressure due to the increase of the electron density, and generally higher Ar ionization. The measured Ti density also has a maximum at the intermediate distance from the target (45 mm) which generally agrees with the other measurements performed in the magnetron discharges [7, 8]. Such a maximum might be related to the depletion of the ground state energy levels found before for DC sputtering [9], and confirmed recently by applying the ROAS technique to a HIPIMS dischrage [10]. On the other hand, the diffusion of the sputtered particles to the reactor walls leads to the further decrease of their densities at large distances from the target.

The time dependences of the measured densities are also different depending on the working pressure (Fig.3). If at 5 mTorr there is a nearly exponential density decay starting at about 50 μs after beginning of the plasma pulse, at 20 mTorr one can observe a wide peak of the density which appears at about 300 μs after the beginning of the plasma pulse. This peak might be attributed to the "wave" of the thermalized species after the HIPIMS plasma pulse [11, 12]. Nevertheless it might also exist solely due to the extensive Penning ionization during the off-time, according to our recent ROAS measurements [10]. Unfortunately, due to rather high intensity of the emission lines during the plasma pulse, LIF measurements were not performed during the pulse.

The ROAS analysis of the absolute densities in the HIPIMS discharge during the plasma off-time performed at 20 mTorr of working pressure and at 5 cm away from the target surface has shown that the zero ground state density of the Ti neutrals in its maximum shown in Fig.3 is as high as $\approx 10^{11}$ cm^{-3}.

The same density of the Ti ions measured by ROAS at 20 mTorr at the same time gives the maximum value of about $\approx 5 \cdot 10^{10}$ cm^{-3} (see Fig.3).

4. Conclusions

The results of LIF measurements of the "short pulse" (10 μs) HIPIMS discharge are presented. The time behavior of the relative Ti and Ti^+ ground state densities as well as the evolution of the corresponding spectral lines broadenings were measured by LIF during the plasma off-time. The measurements were performed at two working pressures (5 and 20 mTorr), and at three distances from the target (15, 45 and 70 mm). The results of this study show that the FWHMs of the spectral lines of sputtered species in the direction parallel to the magnetron target reveal a faster-than-an-exponential decay during the plasma off-time having always a spectral shape close to the Lorentzian one. Based on this, strong non-thermalization as well as strong influence of collisions in the gas after the pulse are suggested. The thermalization trends for Ti neutrals and ions were found to be similar; however, the FWHMs of the Ti^+ spectral lines appear to be generally narrower.

Spatial analysis of the Ti and Ti^+ densities show a slight density increase at the intermediate distance from the target which correlates with previous measurements [8], yet being unclear. The densities measured at 20 mTorr reveal a wide "secondary" peak appearing at about 300 μs after the plasma pulse which might be either due to the wave of thermalized particles [11] or due to an extensive ion production by the metastables species. This result is confirmed by the additional density measurements performed by ROAS [10] at 20 mTorr of working pressure which point out to similar effect.

5. Acknowledgements

The authors would like to thank the financial supports of the Belgian Government through the IAP program (P06/08) and of the Walloon Region (DGO6) through the Opti2mat program. The authors would also like to express their special thanks to Matthieu Michiels and Denis Walrave (Materia Nova research center) for the help related to signal processing. Nikolay Britun is a post doctoral researcher of FNRS, Belgium.

6. References

[1] J. Amorim, G. Baravian, and J. Jolly, J. Phys. D: Appl. Phys. 33 (2000) R51.
[2] L. de Poucques, C. Vitelaru, T. M. Minea, J. Bretagne, and G. Popa, Euro Phys Lett. 82 (2008) 15002.
[3] A. C. G. Mitchell, and M.W. Zemanski, Resonance Radiation and Excited Atoms. Cambridge: Cambridge University Press 1971.
[4] U. Fantz, Plasma Source Sci Technol. 15 (2006) S137.
[5] F. Gaboriau, U. Cvelbar, M. Mozetic, A. Erradi, and B. Rouffet, J. Phys. D: Appl. Phys. 42 (2009) 055204.
[6] Sirah CobraStretch dye laser User Manual, p. "Laser-Description"-3.
[7] N. Britun, M. Gaillard, and J. G. Han, J. Phys. D: Appl. Phys. 41 (2008) 185201.
[8] K. Sasaki, N Nafarizal, J. Phys. D: Appl. Phys. 43 (2010) 124012.
[9] S. Konstantinidis, A. Ricard, M. Ganciu, J. P. Dauchot, C. Ranea, and M. Hecq, J. Appl. Phys. 95 (2004) 2900.
[10] N. Britun, M. Palmucci, S. Konstantinidis, R. Snyders, Proc. 11-th Int. Symp. for Sputtering and Plasma Processes (ISSP), Kyoto 2011, p.198.
[11] S. Konstantinidis, J. P. Dauchot, M. Ganciu, and M. Hecq, Appl. Phys. Lett. 88 (2006) 021501.
[12] C. Vitelaru, L. de Poucques, T. M. Minea, and G. Popa, Jour. Appl. Phys. 109 (2011) 053307.

AUTHOR INDEX

Institute of Physics
Dirac House, Temple Back
Bristol BS1 6BE UK

proceedings.com